A Hierarchical Concept of
Ecosystems

MONOGRAPHS IN POPULATION BIOLOGY
EDITED BY ROBERT M. MAY

A
Hierarchical Concept
of Ecosystems

R. V. O'NEILL
D. L. DeANGELIS
J. B. WAIDE
T.F.H. ALLEN

PRINCETON UNIVERSITY PRESS

PRINCETON, NEW JERSEY

/3526197

11-11-86
DLM

Copyright © 1986 by Princeton University Press
Published by Princeton University Press, 41 William Street,
Princeton, New Jersey 08540
In the United Kingdom: Princeton University Press,
Guildford, Surrey

ALL RIGHTS RESERVED

Library of Congress Cataloging in Publication Data will be
found on the last printed page of this book

ISBN 0-691-08436-X (cloth)
     0-691-08437-8 (pbk.)

This book has been composed in Linotron Baskerville

Clothbound editions of Princeton University Press books
are printed on acid-free paper, and binding materials are
chosen for strength and durability. Paperbacks, although satisfactory
for personal collections, are not usually suitable for library rebinding

Printed in the United States of America by
Princeton University Press, Princeton, New Jersey

Contents

Acknowledgments

Research was funded by the National Science Foundation, Ecosystem Studies Program under Interagency Agreement BSR 8021024 with the United States Department of Energy under Contract No. DE-AC05-840R21400 with Martin Marietta Energy Systems, Inc., and in part by the Office of Research and Development, United States Environmental Protection Agency, under Cooperative Agreement Nos. CR 811060 and CR 807856. Additional funding was provided by Cornell University and by the Office of Health and Environmental Research, United States Department of Energy.

This is publication No. 2719 of the Environmental Sciences Division, Oak Ridge National Laboratory and No. ERC-080 of the Ecosystems Research Center (ERC), Cornell University. The ERC was established in 1980 as a unit of the Center for Environmental Research at Cornell University.

The work and conclusions published herein represent the views of the authors and do not necessarily represent the opinions, policies, or recommendations of the Environmental Protection Agency (EPA) or of Cornell University. The EPA and Cornell University do not endorse any commercial products used in the study.

Part I

The Concept of an Ecosystem

"Ecosystem" is an intuitively appealing concept to most ecologists. Perhaps because of this appeal, the ecosystem concept has become increasingly important since its introduction by A. G. Tansley in 1935. Despite its widespread use, the concept remains diffuse and ambiguous. To many, "ecosystem" has heuristic value but is difficult to define or to associate with any coherent theory or set of principles. The first part of this book will analyze the current ambiguity before proposing our own viewpoint in later chapters.

The approach in Part I may seem overly negative at first sight. Although our critique of past and present concepts about ecosystems will reveal many of their limitations and inadequacies, we do not intend to derogate these concepts or to underrate the tremendous contributions of past and present workers. However, if we are to build new theory, then we must assume that our current ideas are not perfect. In pointing out their weakest points, we may be led to new insights.

Chapter 1 will call into question prevalent concepts of the ecosystem. We will argue that current ideas are fundamentally limited, inadequate as a basis for the development of theory. In particular, the ecosystem concept has often been biased by emphasizing one set of observations over another. We will also contrast the population–community and process–functional schools of thought.

Chapter 2 is a historical review of how ecologists have looked at environmental systems. The review will show that current concepts are rooted in problem areas that have interested ecologists and natural historians in the past. In many respects, current ambiguities are due to limited concepts of time and space that were adopted because they were useful for earlier problems. Once we have identified the cause of the ambiguity we will be in a position to develop a new and more consistent theory.

CHAPTER 1

Fundamental Ambiguities in the Ecosystem Concept

Four blind men are led into a courtyard to experience an elephant for the first time. The first grasps the trunk and declares that elephants are fire hoses. The second touches an ear and maintains that elephants are rugs. The third walks into its side and believes that elephants are a kind of wall. The fourth feels a leg and decides that elephants are pillars.

This well-worn parable contains an important moral for the ecosystem ecologist. Like elephants, ecosystems can be viewed as many perspectives. Our conclusions are biased by the way we observe ecosystems. For example, if we focus on interactions among individual organisms, ecosystems seem relatively constant backgrounds, contexts within which the interesting phenomena occur. If we focus on succession, ecosystems appear to change continuously through time. In fact, both impressions are correct, depending on the purpose and the time–space scale of our observations.

THE ECOSYSTEM AS AN OBJECT OF STUDY

Many biological sciences are fortunate in having the organism as their object of study. Organisms are convenient units, being discrete, of convenient size, and with a life span short enough to be studied as a whole. Of course, none of these properties hold with complete generality. Vegetatively reproducing organisms are not always discrete, many animals and plants are too large for easy study under con-

3

trolled conditions, and the lifetimes of some, such as the bristlecone pine, far exceed the life span of any human investigator. But these exceptions have not diminished the relative ease of regarding the organism as a conceptual unit. In particular, focusing on the organism has permitted the biological sciences to establish a priori the appropriate time and space scale for approaching the object of study.

Ecology has not been as fortunate in having its objects of study so conveniently bounded. The ecosystem has long been recognized as a fundamental organizational unit in ecology (Tansley 1935) and a major structural unit of the biosphere (W. S. Cooper 1926; Krajina 1960). However, many properties of ecosystems, such as boundaries, are abstract (Sjors 1955; Fredericks 1958) and precise definition is far from easy. Evans (1956) argued for using "ecosystem" to describe any organizational unit of interest: organism, population, or community. Dale (1970) defined the ecosystem as any system that was open to the flow of energy or matter and contained at least one living entity.

Ecologists generally recognize ecosystems as a specific level of organization, but questions of appropriate time and space scales remain. Some ecologists see ecosystems as convenient portions of the natural world, arbitrarily chosen for study. Colinvaux (1973) argued that one could choose any size area, provided it had recognizable boundaries. Graham (1925) found it convenient to consider the felled tree as an ecosystem. Fish and Carpenter (1982) dealt with ecosystem processes in individual tree holes. Bosserman (1979) argued lucidly for considering a patch of floating macrophytes as an ecosystem. Other ecologists view ecosystems as the environmental context within which population or community dynamics occur (Hutchinson 1978). Others feel that an explanation of macroscopic properties like nutrient cycling requires a more coherent and integrated conceptualization of ecosystems (O'Neill and Reichle 1980).

Morowitz (1968) proposed that ". . . sustained life under present-day conditions is a property of an ecological system rather than a single organism or species." This logic can be extended to define ecosystems as the smallest units that can sustain life in isolation from all but atmospheric surroundings. However, one is still left with the problem of specifying the area that should be included. A few square meters may seem adequate to a soil microbiologist, but 100 square kilometers may be insufficient if large carnivores are considered (Hutchinson 1978). Unless recognizable boundaries exist (such as the edges around a lake or an island) there is no simple method for specifying the size of an ecosystem. Even in such apparently discrete entities as lakes and islands, imports and exports of organisms and matter might be significant enough to question whether survival in isolation is possible. Of course, it is not necessary to define an ecosystem as the smallest unit that can survive in semi-isolation, but if not, precise definition can become arbitrary. The problem of spatial scaling is vitally important to the ecologist, then, while it is not to the organismal biologist, whose basic unit of study has more obvious bounds.

Since spatial scales are important, temporal scales inevitably are, too. An ecosystem, defined as occupying a certain areal extent (e.g., an island) may be approximately self-sustaining over short periods of time. During short periods, losses of nutrient are negligible and stochastic fluctuations are small, but over longer time scales, such a quasi-equilibrium must inevitably degrade. Good examples are marine islands whose areas have shrunk since Pleistocene glaciation, leaving them supersaturated with respect to their biota. These islands seem to be stable systems only over short time scales.

Thus, the spatial and temporal scales at which one should study ecosystems are not specified a priori as they might be in a higher organism. Ecosystems must be studied with specific spatial and temporal scales in mind.

NEED FOR AN ADEQUATE CONCEPT OF ECOSYSTEMS

It is the task of ecosystem analysis in general and of this book in particular to develop an adequate conception of ecosystems, in order to make meaningful ecosystem comparisons from which general principles may emerge. To achieve integration, we must not focus on any one way of observing the natural world. Instead, we must consider many observation sets. Most importantly, we must address directly the complexity of the natural system that ecology has adopted as its object of study.

The need for an integrative concept of the ecosystem has been indicated by a number of authors (e.g., E. P. Odum 1969, 1977; Overton 1977; Patten 1978), and frequent criticism has been directed at many of the conceptual foundations of ecology (Drury and Nisbet 1971; Regier and Rapport 1978). Ecology has been variously described as a collection of unfalsifiable theories (Murdoch 1966), handicapped by the inadequacy of its conceptual framework (Platt and Denman 1975), more interested in concepts than theories (Rigler 1975), and tautological (Peters 1976). Such criticisms have persisted throughout the history of ecology since its emergence as recognizable discipline (e.g., Cowles 1904; C. C. Adams 1913, 1917). One cannot easily attribute such comments to the relative youth of ecology as a science (T.F.H. Allen and Starr 1982) or to the apparently diffuse nature of its subject matter (McIntosh 1976, 1980).

Pielou (1972) suggested a reason for the current state of uncertainty in ecology. She suggested that ecological phenomena contain a degree of stochasticity that leads to an innate unpredictability. She did not argue that ecological systems are totally unmanageable, only that ecology could

6

never equal the predictive power of physics and chemistry because of the white noise problem. Ulanowicz (1979) provided a similar analysis for the apparent inadequacy of ecosystem models.

However, placing the blame on the stochasticity of ecological phenomena seems to sidestep the real issue. Much of the problem can be handled by careful experimental design. It is far more likely that the failure to deal adequately with the complexity of natural systems has led to the present unsatisfactory state of ecological theory. Ecologists have historically either ignored complexity, circumvented it by working with simple systems, or invented mathematical surrogates for ecosystems.

The most common way to avoid complexity is to overemphasize a single type of observation set. By an observation set we mean a particular way of viewing the natural world. It includes the phenomena of interest, the specific measurements taken, and the techniques used to analyze the data (T.F.H. Allen et al. 1984). The space–time scale is an important property of an observation set, determining the total number of measurements and the intervals between them. For example, one observation set might deal with the number of individuals in each species in an area and a contrasting observation set might contain only the total flux of nutrients into and out of the area. Each of these points of view emphasizes different phenomena and quite different measurements. But since neither encompasses all possible observations, neither can be considered to be more fundamental. When studying a specific problem, the scientist must always focus on a single observation set. However, when developing theory, many observation sets must be considered.

Like the blind men and the elephant, limited views have generated considerable disagreement. Some see ecosystems as linear (Patten 1975), while others present evidence for their nonlinearity (Dwyer and Perez 1983). Some see no basis for a cybernetic view (Engelberg and Boyarsky 1979), while others argue that ecosystems are self-controlling and cybernetic (e.g., Patten and Odum 1981). In many cases, these controversies are caused by jumping from the legitimate statement, "Within some observation sets ecosystems can be conceptualized as though they were cybernetic (or linear)," to the illicit conclusion, "Therefore, ecosystems *are* cybernetic (or linear)."

Probably the most divisive controversy is between the population–community view of ecosystems and the process–functional approach. It will be helpful to look more closely at this dichotomy to illustrate how differences in the way we view the natural world determine our conclusions about its nature.

COMMUNITY VERSUS FUNCTIONAL APPROACHES TO ECOSYSTEMS

The reader will pardon us if, for the sake of presentation, we exaggerate the differences between biotic and functional approaches to the ecosystem. Our purpose is not to accuse either approach of ignorance or naiveté, and we do not wish to decide on the superiority of either. By sharpening the distinction, we wish to illustrate that focusing on only one aspect of a complex system can lead to very different views on the nature of ecosystems.

The Population–Community Approach

Population–community ecologists tend to view ecosystems as networks of interacting populations. The biota *are* the ecosystem and abiotic components such as soil or sediments are external influences. The biota may interact with the abiotic environment, but the environment is largely viewed as the backdrop or context within which biotic interactions occur.

It is not surprising that ecologists should focus on the species population. Weiss (1971) suggested that human evolutionary history led to the recognition of conspicuous and tangible entities as suitable objects of scientific inquiry. Early human survival depended on the ability to perceive discrete organisms and classify them as dangerous or nutritious. It is not a very large jump from recognizing discrete objects to using them to characterize the natural world. But notice that it is the limited spatiotemporal framework of the human sensory system that leads to this conclusion, not any inherent property of the intact landscape. From the human scale, it is natural to assume that a list of constituent species, weighted by relative abundance or biomass, constitutes a fundamental description of an ecosystem.

This viewpoint is reinforced by the clearly defined problems that can be approached on this scale: predation, competition, population growth, and so on. This viewpoint is also encouraged by the availability of an explanatory tool, the theory of natural selection (e.g., Huxley 1943). Natural selection operates on the scale of the population and provides a powerful unifying theory.

The Process–Functional Approach

In proposing the term "ecosystem," Tansley (1935) recognized the problems inherent in focusing exclusively on biotic entities. To him, the fundamental unit included both organisms and physical components. Though some problems require us to focus on the biota, other questions require that organisms and their physical environment be considered as a single integral system. Lindeman (1942) later echoed Tansley's viewpoint by defining the ecosystem as ". . . the system composed of physical-chemical-biological processes active within a space–time unit."

E. P. Odum (1953) built upon Lindeman's work and he is largely responsible for developing the process–functional approach. He emphasized energetics as the central focus of ecosystem inquiry (E. P. Odum 1960). A related view (e.g., Pomeroy 1970; Waide et al. 1974) can be traced back to Lotka (1925) and emphasizes biogeochemistry as a point of entry to the analysis of ecosystem function. Because of the orientation to process, the functional approach is able to deal with cyclic causal pathways (Hutchinson 1948) that are often unobservable in population–community problems but which form feedbacks essential to ecosystem maintenance. In an extreme form, the functional approach implies that energy flow and nutrient cycling are somehow more important or more fundamental than the biotic entities performing the function.

The process–functional approach deals with a range of spatiotemporal scales. At one extreme, functional studies emphasize measurement of inputs and outputs of total landscape units (Bormann and Likens 1967; Likens et al. 1970). At the other extreme, the approach seems appropriate for dealing with processes such as decomposition. Here biotic and abiotic components are tightly linked and it is difficult to externalize the nonliving dynamic components. In addition, it is practically impossible to isolate individual species populations of bacteria and fungi.

Consequences of the Difference in Approach

Although few ecologists would hold to either extreme of the spectrum, most will be drawn in one direction or the other by the specific problems that interest them. The two schools of thought focus on different observation sets, and different concepts of the ecosystem are required to explain their observations. They differ, for example, in the most convenient way to subdivide ecosystems into components. To illustrate the difference, Figure 1.1 compares two

BIOTIC EMPHASIS   FUNCTIONAL EMPHASIS

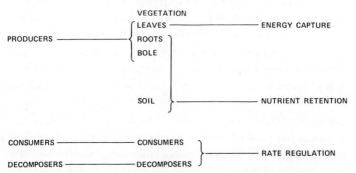

Fig. 1.1. Differences in ecosystem groupings according to biotic and functional viewpoints. The functional subdivision is based on O'Neill 1976.

groupings of ecosystem components, each attempting to describe energy flow.

A typical subdivision results in primary producers, consumers, and decomposers. This subdivision does not ignore function but is designed for problems dealing with a species population and its contributions to ecosystem function. The system is conceptualized in terms of aggregates of biotic units. A biotic interest is clear in that soil organic matter is often considered as external to the system. Of course, the soil is not ignored, but it is viewed as part of the environment, operating on the system from outside.

A second subdivision (O'Neill 1976) emphasizes energy capture, nutrient retention, and rate regulation. The components of the system are more directly functional. Traditional trophic levels (i.e., consumers and decomposers) are combined on the basis of their hypothesized role of regulating rates within the system. Only rapidly metabolizing portions of the plants (e.g., leaves) are considered to have a role in energy capture. Interestingly, the nutrient retention component is composed of part of the plants (boles and

11

large roots) and part of the physical environment (soil organic matter).

Although still relying on lumping distinct biological entities, O'Neill's grouping is based primarily on functional roles in the system. The nutrient retention component is specifically defined on the basis of rate processes. To define the components on this basis, it is necessary to discard the a priori human scale of perception. The trees do not belong to a single subsystem. The leaves are assigned to one subsystem and the boles to another. The analysis illustrates that a functional view of ecosystems need not be based on species or organisms. However, this conception would be quite inconvenient for a problem dealing with the loss or addition of any single population and the subsequent effect on ecosystem function.

The distinction we have drawn between biotic and functional views has been exaggerated for the sake of presentation. For some ecologists, ecosystems are either biotic assemblages or functional systems. For most, the ecosystem concept contains elements of both. Textbooks (e.g., Collier et al. 1973; Pianka 1974; R. L. Smith 1966) tend to follow E. P. Odum (1971) in defining ecosystems in terms of interacting biotic and abiotic components forming a functional unit. But though we have made the distinction overly sharp, it remains true that much of the ecological literature can be neatly classified into one category or the other (see discussions in Levin 1976). We do not possess an integrative concept of ecosystems which gives equal play to both viewpoints. Outgoing presidents of both British (MacFadyen 1975) and American (F. E. Smith 1975) ecological societies have emphasized the need to draw these viewpoints more closely together.

From one perspective, it can be maintained that the difference between the viewpoints is nothing but a methodological question. Whenever one approaches a complex system, one is wise to focus on a specific aspect of the system (e.g., either population interactions or nutrient cycling). The functional ecologist is interested in energy and material processing, while the community ecologist focuses on population interactions. The progress of ecology demonstrates that each viewpoint is valuable and that there are clearly defined problems that can be profitably approached by each.

It is not this methodological question that concerns us here. It is clear that both methods are valuable, but it is also clear that extreme proponents of each viewpoint seem to consider theirs as somehow more fundamental. It is this controversy over the nature of ecosystems that is our primary concern. The viewpoints are often seen as contradictory rather than complementary.

To illustrate the controversies that continually arise, we can look at recent discussions of whether or not ecosystems are cybernetic. Engelberg and Boyarsky (1979) argued that ecosystems (seen as complexes of interacting organisms) lack a communication network capable of linking all of the parts into an integrated system. The biota lack informational linkages designed to "steer or regulate" the system. The authors imply that to deal with ecosystems as systems requires an overextension of the organism concept.

However, Quinlin (1975) argued that a cybernetic view of ecosystems is appropriate. She maintained that ecosystems (seen as a complex of interacting functional components) possess the relevant informational links in the form of persistent patterns of structure and function which coordinate nutrient cycling through a concomitant dissipation of energy. Innis (1978) has argued for the usefulness of the cybernetic paradigm for nutrient cycling studies. Patten and Odum (1981) also argued that the cybernetic concept was applicable to ecosystems when one considers the process–functional viewpoint.

13

THE LIMITATIONS OF PARTIAL VIEWPOINTS

What emerges is that neither the population–community nor the process–functional viewpoint can serve as the exclusive or fundamental concept of ecosystems. Each chooses a different observation set with clear advantages for specific problem areas. But we must now also consider the ways in which each viewpoint is critically limited so that neither can be considered as *the* way to see the natural world.

Limitations of the Process–Functional Viewpoint

The functional or process viewpoint displays its fundamental limitation whenever it looks at fluxes of energy, material, and information as though they existed independently of the species involved in them. For many, this suggests an unacceptable form of holism that exaggerates the analogy between the organism and the ecosystem. Indeed, in the past, many authors have written as though the community or ecosystem *was* an organism (e.g., Taylor 1935b; Carpenter 1939; Allee et al. 1949).

A set of experiments has been carried out that gives considerable insight into the limitations of a purely functional approach. Three tests were designed for remote execution by the Viking I and II Martian landers, each involving material and/or energy fluxes in the Martian soil (H.S.F. Cooper 1976). It was thought that these tests would determine whether or not microbial life exists on the surface of that planet.

In the first test, a nutrient broth was added to Martian soil. If heterotrophs were present, a gas chromatograph should detect carbon dioxide, oxygen, hydrogen sulfide, et cetera. When the broth was added, an immediate burst of oxygen was recorded, about twenty times what would be expected if no life were present.

The second experiment added a nutrient solution with

14

carbon-14-labeled carbohydrates and amino acids to Martian soil. Heterotrophic activity should produce labeled carbon dioxide, measured by a radiation detector. In the experiment, radioactivity was immediately detected at a rate faster than would have been produced by the richest organic soils on Earth under similar circumstances.

The third experiment was aimed at detecting autotrophs. Soil was placed in a chamber with simulated sunlight and carbon-14-labeled carbon dioxide and carbon monoxide. Later, the atmosphere was flushed and the soil heated to break down the microbes. Organic compounds of greater molecular weight than methane were captured in an organic vapor trap and counted with a radiation detector. In the analysis, a count of more than six times what would have been expected from sterile soil was recorded.

The day-to-day drama that surrounded the reception and interpretation of these Martian lander data is described in a lively manner by H.S.F. Cooper (1976). The case for life on Mars looked very strong at times, but the accumulated weight of evidence and opinion came down against the hypothesis of life. Experiments on Earth demonstrate that an oxidant, such as hematite, could have caused the same measurements. In the first experiment, the addition of water could have split oxygen atoms from the oxidant. In the second, the released oxygen might have combined with carbon-14 to produce the observed carbon dioxide. In the third experiment, a complex molecule (e.g., formic acid) could have formed through ultraviolet light-stimulated reactions.

The story of the search for life on Mars contains an implicit warning. The experiments took a purely functional approach and made no attempt to demonstrate that an *organism* was performing these functions. To demonstrate life, a microscopic examination for living organisms would have been required. Ecosystems are not merely the sum of material and energy flows. These fluxes are indispensable but they constitute only one aspect of the system. If we attempt to deal with ecosystems strictly on the basis of flows and fluxes, we run the risk of ignoring essential features of biological organization.

The process–functional approach is limited in areas that deal with the effect of single populations on ecosystem function. In some ecosystems, a dominant or key species (see Chapter 9) will strongly influence overall system function. The loss or addition of populations such as forest trees or bison significantly alters process rates. Because species changes are not directly considered in the functional approach, such situations are difficult to represent.

Another problem with applying a purely functional concept is that ecosystems must be conceptualized differently for different chemical elements. Each element has unique chemical properties and displays distinctive biogeochemical behavior. This is especially true of elements with a sedimentary cycle (e.g., calcium) as opposed to ones with a gaseous cycle (e.g., nitrogen). Thus, nitrogen cycling requires consideration of nitrogen-fixation and denitrification, processes that do not occur in the calcium cycle. The ecosystem appears to have distinctively different structural properties when examined for different elements.

Limitations of the Population–Community Approach

A community, like an ecosystem, requires some degree of abstraction before it can become an object of study. There does not exist a set of phenomena forced upon us by the natural world that requires us to set up communities to explain them. Instead, the community is operationally defined for the researcher with a particular taxonomic background and a particular suite of sampling methodologies.

In part, the community viewpoint is a result of the historical development of ecology. In recent decades, ecological theory has developed largely around observation of vertebrates and higher plants. These are the organisms that correspond most closely to our human perceptual scale, and it is natural that we should be most interested in them. But at the same time, it is not legitimate to assume that this convenient scale of observation will necessarily be useful for all problems.

Mattson and Addy (1975) point out that consideration of the plant community as an entity separate from its animal community created problems in the study of plant–insect interactions. Earlier, Drury and Nisbet (1971) made a similar point, showing how isolation of organisms from their biotic and abiotic environment has complicated the understanding of plant–herbivore dynamics. When a plant is considered as a part of the animal's environment, as a sort of structure in its habitat, interactions become difficult to conceptualize. Thus, by including plants in the environment of large herbivores, current theory underestimates the interaction of plants and animals in the evolution of grassland species (Major 1969).

The concept of a separate environment is particularly troublesome when it is combined with a species orientation. In this perspective, entities that do not fit the species or population concept would have to be considered as "outside" the system (i.e., in the environment). Thus, if one tried to apply this conception to a nutrient cycling problem, one would be left with the impossible situation of conceptualizing a nutrient atom as leaving the ecosystem in litterfall, reentering the system (perhaps several times) as it is consumed by decomposers, leaving the system with the death of the decomposer, and reentering it as it is absorbed by a plant. In this awkward situation, there is no cycling at all. There is simply a series of entries and exits.

The community concept is also awkward when one is dealing with belowground processes. Here the taxonomy is

17

poorly defined and phenomena do not fall out naturally from measurements made on individuals. Here functional components that do not fit easily into the community concept begin to force themselves on us. Consider, for example, a problem that deals with the rhizosphere. In this case, the functional component is composed of a part (root hairs) of one organism, soil water, clay particles, and a complex of microorganisms. There is seldom a simple relationship between species and functional components (J. E. Schindler et al. 1980).

It is clear in some cases that there is considerable redundancy of function among species populations. We know, for example, that primary production in a planktonic system is often determined by limiting nutrients. Given the same inputs, the same level of production occurs no matter what species of phytoplankton are involved (Vollenweider 1975, 1976; D. W. Schindler 1977, 1978). Species act as compensating devices (Whittaker and Woodwell 1972), and ecosystem function may be insensitive to a range of variability in species abundance. Thus, with important exceptions such as indicator or key species (see Chapter 9), the ability to infer ecosystem properties directly from species properties is limited.

Thus, the community approach is an appropriate conceptualization for some observation sets, rather than the best or most fundamental way to view ecosystems. There are other observation sets for which the community approach is awkward at best. The community viewpoint has as much to do with the way we look at ecosystems as it has to do with the natural world itself.

Summary

Some degree of abstraction is required in order to study ecological systems. Unlike those biological sciences that focus exclusively on the organism, ecology cannot set up a single spatiotemporal scale that will be adequate for all investigations. As a result, ecologists must be careful not to extrapolate from any single type of observation to the nature of the underlying system. Lack of care in this regard is largely responsible for the ambiguities surrounding the ecosystem concept.

It becomes clear that neither the process–functional approach nor the population–community approach can form a complete theoretical foundation for ecosystem analysis. Like the blind men and the elephant, both approaches adopt limited viewpoints. To find an integrative concept, we must pay careful attention to the different ways in which ecosystems can be viewed and how one's perspective changes the conception needed to explain the observations.

In this regard, it will be particularly helpful to survey the ways that ecosystems have been thought of in the past. To this end, the next chapter will take a historical perspective. In particular, we will be concerned with how the concept of the ecosystem changed as different problems, across a range of spatiotemporal scales, were addressed.

CHAPTER 2

A Historical Perspective on How Ecologists Have Viewed Ecosystems

Many of the problems outlined in Chapter 1 can be understood as the natural consequence of the way ecologists have viewed the natural world. Over time, different concepts about ecosystems have dominated as attention was focused on one problem area or another. Documenting this alternation is the first objective of this chapter.

Our second objective will be to illustrate how emphasis on only one type of observation or another can lead to quite different conceptions of the natural world. Depending on the observation set, ecosystems have been seen as static or dynamic, as steady-state or as fluctuating, as integrated systems or as collections of individuals. Often we will find that success in solving one limited set of problems led to the erroneous conclusion that a similar approach would be useful in all other problem areas.

Our third objective is to focus on an important characteristic of the observation set, its spatiotemporal scale. As problems change, the scale of observation changes appropriately. Ecological principles often do not translate well across these scales and much confusion has resulted. In later sections of the book, we will find that explicit consideration of scale is critical to developing an adequate concept of ecosystems.

STATIC VERSUS DYNAMIC VIEWPOINTS

Biogeography, Phytosociology, and the Static View

A major question for naturalists of the eighteenth and nineteenth centuries was pattern in the geographic distribution of organisms (Colinvaux 1973). Early plant geographers (e.g., Humboldt 1807; Candolle 1874; Warming 1895) described major plant formations distributed according to macroclimatic factors, especially temperature and moisture. Vegetation maps prepared by these naturalists were translated into climatic maps and influenced the interpretation of global climatic patterns (Koppen 1884). Raunkiaer (1934) developed life-form spectra for the major formations, based on location of the perennating organs of plants. Dokuchaev (1883) later interpreted the distribution of major soil types in relation to vegetation and climate. On a smaller scale, Merriam (1890) recognized the zonation of plants and animals on the San Francisco Mountains and later expanded his life-zone concept (Merriam 1898) to the entire North American continent under the influence of J. A. Allen's (1871) earlier work.

The dominant view that emerged from this activity was that global patterns of atmospheric circulation partitioned the earth into distinct climatic zones. These zones presented common adaptive challenges to organisms, resulting in the observed patterns of vegetation. Climate and vegetation influenced soils, which, in turn, affected the vegetation. This view of strong interdependencies among climate, biota, and soil emphasized the long-term stability of the present pattern. The pattern would not change without a major change in climate. The resulting view of the natural world was a static one that deemphasized any short-term changes in the system.

This static view was intensified by subsequent developments in phytosociology (see Whittaker 1962 for a discussion). The Zurich-Montpellier school, under the leadership of Braun-Blanquet (1932), identified plant associations on the basis of the fidelity of characteristic species. Observations were taken in areas judged by the researcher to be typical of the plant association. Proponents of the Uppsala school, under the tutelage of Du Rietz (1929), classified communities on the basis of species constancy in randomly located quadrants. Both approaches were highly typological and led to a classification scheme not unlike Linnaean taxonomy. A similar approach to community classification was adopted in North America, based on the concept of dominance (Cain and de Castro 1959).

The phytosociological approach has been very influential and continues to generate controversy (Mueller-Dumbois and Ellenberg 1974; Shimwell 1971). Its success in Western Europe is probably due to the disjunct distribution of vegetation resulting from glaciation and the small scales resulting from centuries of human disruption. Because the vegetation cover was not continuous, gradations between types were not conspicuous. It was easy to visualize the system as a set of discrete and constant vegetation types.

Within the spatiotemporal scales adopted in these studies (i.e., within these observation sets), the natural world appeared constant. Although biogeographers often considered successional changes in their discussions, short-term dynamics were a minor part of their overall conceptualization. Small-scale variations were ordinarily smoothed over broad geographic areas to find a constancy beyond the perceptual scale of a single human investigator. Problems resulted whenever this limited vision was taken as evidence that the natural world was "truly" or "only" static in character.

Succession, Paleoecology, and the Dynamic View

Explicit consideration of temporal dynamics began at the turn of the century with the emergence of the concept of succession. Cowles (1899, 1901) pioneered studies on the Michigan sand dune succession and brought the concept to the forefront of ecological attention. But it was Clements (1916, 1936) who codified succession and developed it into a dominant ecological theory.

Clements envisioned plant communities as developing over time to a common, stable endpoint, the climax association, determined by macroclimate over a broad region. The influence of earlier biogeographical studies on Clements is obvious. Braun (1950) was influenced by Clements's ideas as she developed a model of the development and organization of the deciduous forest of eastern North America.

The introduction of the concept of succession was extremely important to the development of ecology and helped broaden the ecologist's perception of time. But, given the emphasis on a uniform and stable climax, the viewpoint was still limited. A. S. Watt (1947) later commented that much more would be needed to understand the dynamics of the plant community or to formulate laws for its maintenance and regeneration.

The problem areas being addressed did not require linking succession with processes operating over other temporal scales. Successional time is not the only, nor even the most important, scale for ecological organization. It is likely that succession has played a central role only because man can perceive the progressive changes that occur during the first years following disturbance. Indeed, it is widely recognized that succession and climax, as used by terrestrial ecologists, do not extend easily to aquatic planktonic systems (Whittaker and Woodwell 1972). Planktonic systems operate on a very different spatiotemporal scale, yet we clearly want to deal with them as ecosystems. Therefore, the ecosystem concept cannot be locked into a successional time frame.

The emerging temporal perspective was reinforced by

studies of paleoecology in the early part of this century. Analysis of plant remains in Scandinavian peat bogs delineated six climatic epochs in postglacial time (Sernander 1908). Independent evidence for long-term changes was provided by the development of pollen analysis by von Post in 1916 (Colinvaux 1973). Pollen analysis showed that the changes were regional in character, rather than local. Close correspondences in pollen analyses were obtained throughout northern Europe. The results suggested eight rather than six major climatic episodes, all characterized by consistent differences in vegetation composition. These paleontological studies strengthened the view that plant formations existed over broad geographic regions. The formations were relatively constant in space and changed only when the climate changed. The perspective of temporal dynamics had been expanded only very slightly.

Animal Demography and Ecosystems as Equilibrium Systems

While plant ecologists worked on large spatial scales and considered slow temporal changes, another group of ecologists focused on theories of population growth and regulation (Lotka 1925; Volterra 1926; Gause 1934). These theories couched population processes in the language of Newtonian–Laplacean mathematics, and time was an explicit variable in the differential equations.

These studies paid explicit attention to temporal dynamics. Early models considered the temporal pattern of population growth (e.g., logistic growth) and were concerned with explaining temporal cycles in population size. However, the full complexity of ecological time and space scales was not yet addressed. For example, the nonlinear population models did not yield to analytical solution, and this led many investigators to emphasize the steady-state properties of the equations. Thus, although these formalisms explicitly considered dynamics, they may have actually intensified

24

the prevailing static viewpoint by emphasizing steady-state or equilibrium conditions. This emphasis generated numerous controversies about the regulation of population size (e.g., Cole 1957) and introduced new static concepts such as the equilibrium community of Hutchinson (1953). Such debates emphasize the limitations to the concept of constancy within which ecological theories were being constructed. Controversies about population regulation boiled down to what mechanisms are operating to keep the populations constant. Indeed, theories of population regulation can be seen as formalizations of the "balance of nature" concept. This equilibrium concept has influenced biologists and naturalists since antiquity (Egerton 1973) and has been a key element of Western intellectual tradition since the modern scientific revolution.

The concept of the balance of nature, termed by Egerton (1976) the oldest ecological theory, represents man's concept of the environment as viewed from his own spatiotemporal framework. Changes in the landscape occur very slowly relative to the daily and seasonal changes that determine the pace of human events. The changes are slow even relative to a human's life span. Therefore, we tend to view the environment as constant and seek concepts that help explain how it maintains this constancy.

Clementsian versus Gleasonian Views of the Natural World

Clements's (1916) view of succession stimulated a great deal of interest in the dynamics of plant communities. Clements saw the community as a distinct dynamic entity that reacted to environmental gradients as a unit and was integrated by interactions among component populations. The Clementsian view of the community was largely developed in botanical studies but also appears in the work of Shelford (1913) and Elton (1927). The concept was extended by Allee and his colleagues at Chicago (Allee and Park 1939; Allee et al. 1949).

Phillips (1931, 1934, 1935) defended Clements's views and generated considerable confusion by playing heavily on the analogy between organism and community. Phillips's concepts were derived from holism and emergent evolution presented by Smuts (1926). Developed from creative evolution and from process philosophers such as Whitehead, Smuts's naive holism was largely unacceptable to a scientific audience. Smuts maintained that the emergence of properties at new levels of organization, such as the community, required special explanations. He posited that emergent properties required a kind of "soul," an *élan vital*, an entelechy. The analogy between the organism and community was being overdrawn.

Reaction to the superorganism concept was quick in coming. Gleason (1917, 1926, 1939) stressed individualistic behaviors of single species in response to environmental gradients. Gleason saw the community as an aggregate of species populations with similar tolerance to environmental conditions. Where others saw a supraorganism, Gleason saw individual species populations in continuously changing environments.

The focus on species properties has a long history of development in zoogeography and animal ecology. The emergence of Darwinian theory led nineteenth-century zoogeographers to split the globe into regions emphasizing common ancestries and evolutionary affinities. Biogeography focused on species and their evolutionary and taxonomic relationships, rather than on communities having similar distributions in response to major climatic patterns. Developments in animal demography tended to reinforce the individualistic view. The mathematical theory of populations combined nicely with evolutionary concepts and Mendelian genetics (Fisher 1930) into a powerful synthesis.

26

The success of the approach encouraged work on the level of the population.

The controversy between Clementsian and Gleasonian views of the community reminds us of the dichotomy between population–community and functional–process views of the ecosystem (Chapter 1). Both controversies have a similar source, since both result from adopting a limited view of the natural world. Clementsians, like functional ecologists, focused on the integrated system and deemphasized observation sets that did not directly relate to the integrated system concept. Gleasonians, like population–community ecologists, focused on the species and deemphasized the properties that emerge from interactions. It is incorrect to trace the current ambiguities in the ecosystem concept directly to Clements and Gleason, but resemblances in the controversies are unmistakeable.

LINKING SPACE AND TIME

Clements stressed temporal dynamics but deemphasized spatial pattern. Discrepancies between the predicted climax state and the actual pattern of vegetation on the landscape generated a complex and convoluted set of explanatory terms (Whittaker 1953). Gleason argued that spatially heterogeneous patterns were important and should not be dismissed as exceptions to the rule. Spatial patterns were real and should be interpreted in terms of individualistic responses to more or less continuous spatial gradients. Thus, Gleason stressed the spatial pattern but paid less attention to temporal concerns. Neither view arrived at an integrated concept of the linkage between space and time.

Further investigations in paleoecology suggested that dynamics in time and space could be coupled. Piston-driven sample cores (Livingstone 1955) extended studies to lake sediments, less influenced by local input of pollen from bog plants. Radiocarbon methods (Libby 1955) permitted independent dating of samples, thereby freeing palynology from a strictly stratigraphic approach.

Braun (1950) had argued for the constancy of the southern deciduous forests. She felt that south of the furthest advance of glaciers the current spatial distribution had been relatively unaltered by either the glacial advance or concomitant climatic changes. But new pollen analyses from a variety of southeastern sites (Frey 1953; Whitehead 1964, 1965) showed convincingly that the southern forests were profoundly changed by glaciation. In addition, postglacial migration patterns involved individual species rather than movement of intact vegetation units (Whitehead 1965; Davis 1976). Apparently, Braun had been so influenced by Clementsian ideas that she did not conceive of the forest as changing spatially in response to temporal changes in climate. The new pollen analysis techniques supported the perception that spatial dynamics occur in a temporal context, but still there was no explicit attempt to link the two aspects of behavior.

During the same period, Curtis (Curtis and McIntosh 1951; Bray and Curtis 1957; Curtis 1959), Whittaker (1956, 1967), and Goodall (1954) developed gradient analysis to investigate spatial patterns. This approach confirmed the results of the pollen analyses. Species distribution patterns are more in accord with the views of Gleason than with those of Clements. The evidence supported the principles of species individuality and community continuity elaborated by Gleason (1926) and Ramensky (1924). Species are distributed continuously along environmental gradients, leading to continuous intergradation among recognizable plant associations. Abrupt discontinuities in vegetation are associated with abrupt discontinuities in physical environment (Whittaker 1975). Once again, spatial dynamics were

28

emphasized and there was little need to link changes in space and time into an overall perspective.

Whittaker's Attempt to Link Space and Time

Whittaker (1953) combined gradient analysis and succession theory to produce the "climax pattern hypothesis." The climax is not a fixed entity controlled by regional climate. It is a localized intersection of species responding to complex and interrelated gradients in the physical environment. The climax stage is recognized only generally as an approximate steady state of species turnover, biomass, organic matter production, decomposition, and nutrient circulation (Whittaker 1975). The climax pattern hypothesis couples Clementsian dynamics in time with Gleasonian dynamics in space. It recognizes that observed patterns represent the distribution in space of processes operating in time. The spatial patterns may appear continuous or discontinuous, depending on the relative steepness of the gradients and on the spatiotemporal perspective of the human observer.

In some sense, Whittaker's hypothesis represents the culmination of the events chronicled here. Dynamics in space and time were finally related to each other. However, the hypothesis was never intended to be a unified theory of spatiotemporal relationships and is incomplete on two counts.

First, there is no explicit consideration of the spatial coordinates for processes acting at a wide range of temporal frequencies. The temporal perspective is restricted to successional time. No effort is made to relate spatial dynamics to processes acting over very short time scales.

Second, Whittaker's hypothesis applies primarily to observation sets that focus on species individuality. Focusing on similar observation sets, Gleason (1936) referred to the plant association as a "fortuitous juxtaposition of plant in-

29

dividuals." The individualistic concept deemphasizes other determining forces in order to elucidate Darwinian natural selection operating at the population level.

While such a viewpoint is appropriate for some observation sets, it is less useful for explaining other observable phenomena. For example, recycling of nitrogen occurs in a wide variety of plant communities, irrespective of the species composition. In general, a problem results if the individualistic concept is taken to an extreme by concluding that the absence of demonstrable, discrete, spatial groupings of species implies the absence of higher-level organization of any kind.

We must seek beyond the climax pattern hypothesis for an adequate spatiotemporal framework. Other levels of organization must be considered or else the scope of ecological investigations becomes constrained within very limited observation sets. In fact, processes from predation to nutrient cycling occur across a broad spectrum of time and space scales and are associated with different levels of ecological organization. The central features of organization are only definable by objective reference to a range of temporal scales, each acting at a related spatial scale. Explicit consideration of multiple scales is essential for addressing the complexity of ecological phenomena. Inability to deal with the complexity of the spatiotemporal framework of ecological systems is one of the underlying causes of the current ambiguity in the ecosystem concept.

The Pattern–Process Hypothesis

A classic paper by A. S. Watt (1947) represents the next logical stage of development, even though it antedated and influenced Whittaker's work. Watt presents perhaps the clearest available discussion of the complexity and interdependence of ecological scales. He early (1925) recognized the inadequacy of the prevailing spatiotemporal paradigms and formulated a revised concept of vegetation pattern (in space) and process (in time). Using data from seven communities in the British Isles, Watt argued cogently for viewing community dynamics as a cyclical upgrade–downgrade process. At each point in space, there occurs a sequence of phases following one another regularly over time. Watt's phases correspond to the vegetational stages in succession. The whole temporal progression is distributed as a pattern of patches across the landscape. The orderly sequence of phases at each point in space accounts for the persistence of the overall pattern. Moreover, completion of the sequence at any point depends not only on that patch by itself, but on the spatial pattern, because other patches serve as seed sources. We hasten to add that this was a concern of Gleason (1926) and was not overlooked by Clements (1916).

Watt identified the patches in space and phases in time on the basis of relative species abundances. However, he also realized that species composition was highly correlated with other variables such as litter standing crop, soil organic pools, and altered microenvironments. Thus, he argued for the plant community as an integrated sequence of diverse phases forming a space–time pattern. What was constant was the complex pattern across the landscape. Without changes through time at each point and without differences among patterns, there was no constancy at all. There was no organization without complexity in time and space.

Watt's analysis represented three contributions to the ecologist's perception of spatiotemporal frameworks. First, a wider variety of scales is considered. Succession was subsumed in his "upgrade" concept, the orderly sequence of phases at a site. He also considered time and space scales at which the total landscape changed.

The second advance involved the explicit coupling of dynamics in space and time. The succession of phases in time was distributed in space, and different spatial patches were out of temporal sequence because of local history. Indeed, the whole distribution of patches was changing over time. Different spatial scales required consideration of different temporal scales. The two aspects of the dynamics were closely related.

Watt's third advance resulted from his perception that the interrelated dynamics in space–time defined a higher level of ecological organization, the plant community. The whole mosaic of patches, changing in time, was the plant community. He argued that an understanding of the community required an understanding of the relationship of the phases to each other. Thus, he differed with the extreme interpretation of species individuality by emphasizing the objective existence of communities. He felt that Gleason minimized the significance of the interaction among the components of the community. Watt considered these interactions as a primary bond in the maintenance of the integrity of the plant community. Different patches provide seed sources required for the sequence of phases. Different phases change the microenvironment and alter its suitability for other species. It is precisely these interactions that make the plant community a coordinated system.

Watt's analysis was visionary. Although elements of his analysis can be found in earlier (and later) studies, he demonstrated a unique sensitivity to the question of complex spatiotemporal scales. But even Watt's analysis was specific to his observation sets and did not encompass the full range of spatiotemporal scales that characterize ecological phenomena. For example, he did not consider a general framework applicable to both terrestrial and aquatic systems.

In spite of the relevance of Watt's paper to current developments in ecology, it apparently had little impact on ecological theory during the intervening years. Loucks (1970) examined the compositional dynamics of Wisconsin forests and detected wave-form phenomena probably triggered by randomly occurring forest fires. Loucks's analysis clearly provided evidence in support of Watt's pattern–process hypothesis. However, Loucks discussed his results in the framework of gradient analysis and did not explore the spatiotemporal implications of his findings.

More recently, Bormann and Likens (1979a, b) have essentially revised Watt's pattern–process hypothesis. By analyzing data on a variety of possible disturbances such as fire and wind storms, they concluded that steady-states were commonly reached in northern hardwood forests. They proposed a model, the shifting mosaic steady-state, to account for the phenomena. Regional forest dynamics result from individual plot transitions among three organizational states defined by total biomass. Steady-state occurs when the proportions of the total forest in the three biomass states reach approximately constant values. The proportions are set by the prevailing spatiotemporal pattern of endogenous (e.g., death of large canopy trees) and exogenous (e.g., forest fires) disturbances typical of the region.

The intent of Bormann and Likens's analysis was to demonstrate a steady-state for forest dynamics, not to develop a theoretical spatiotemporal framework for ecosystem dynamics. In order to integrate over broad geographic areas, they paid less attention to organizational states that may be demonstrable at smaller scales. Nevertheless, it is clear that their hypothesis is related to Watt's point of view.

CONCLUSIONS

What we have tried to show with our historical survey is an increasing awareness among ecologists of the complexity of the natural world. The prevailing point of view evolved from spatiotemporal constancy to coupled dynamics in space and time that define different levels of ecological organization. This evolution is continuing. T.F.H. Allen and

Starr (1982) refer to a groundswell of interest in questions of ecological scale. Platt and Denman (1975) noted an increased emphasis on periodic behavior defined in both space and time. Levandowsky and White (1977) reviewed a wide literature in demonstrating that ecological processes operate at distinct time–space scales. Current concepts of stream ecosystems explicitly link space and time through partial differential equations (e.g., Newbold et al. 1982).

In spite of the increased awareness of space–time dynamics, no general theory yet exists that accounts for such phenomena in a comprehensive fashion. We continue to extrapolate to the nature of the ecological system from one limited viewpoint or another. These partial approaches have resulted in many of the ambiguities pointed out in Chapter 1. Now that we understand the historical roots of our problem, we must strive to free ourselves from their limitations as we pursue a more comprehensive theory in the subsequent chapters.

Part II

Preliminary Considerations

Part I (Chapters 1 and 2) examined current ecosystem concepts and explored their historical roots. Our analysis showed that overemphasizing any single observation set limits the development of theory. Choosing large spatial scales showed ecosystems as constant or in equilibrium. Emphasizing either the individual population or total system function made it difficult to deal with all possible observation sets. Limited viewpoints led to limited theory.

In Part II, we will present the concept of a "system" (Chapter 3) that is "hierarchical" in organization (Chapter 4) as a good foundation on which to build ecosystem theory. However, both of these terms have been used extensively and have a number of questionable associations. Therefore, we will have to maintain a critical approach to be certain our theoretical foundation is sound. We must be sure what we do *not* mean by a hierarchical system before we can develop theory.

Chapter 3 will introduce the concept of a complex system. We will examine different types of complexity and will characterize ecological systems as having "organized complexity." We will then consider whether ecosystem organization should be characterized as cybernetic.

Chapter 4 will examine the concept of hierarchy. We will be particularly concerned with how the concept is used in ecological literature and in biology in general. We will find that the typical biological hierarchy (i.e., cell, organism, population, community, ecosystem) will not be helpful in developing ecosystem theory. We will argue the need for a more precise definition of hierarchy, which will be developed in Part III of the book.

CHAPTER 3

The Ecosystem as a System

We can observe the natural world at many levels of resolution. Ecosystems are not simply spatially disjunct groupings of taxa (e.g., the plant community). Ecosystems cannot be arbitrarily assigned to a preconceived spatiotemporal framework (e.g., the climatic climax). Ecosystems cannot be conceptualized simply as functional entities or simply as collections of species.

Instead, ecosystems must be viewed as systems in their own right. Phenomena must be measured and explained at the relevant spatiotemporal scale. There is no reason to assume that concepts, theories, or components, defined on some other scale of resolution, will necessarily be applicable to ecosystem phenomena. Stommel (1963) showed that translating concepts across scales can lead to real problems in studying oceanographic phenomena. Similarly, Gould (1980) argued that evolutionary principles cannot be extrapolated across scales (e.g., from the individual to the species and from microevolution to macroevolution). Indeed, Tansley (1935) clearly perceived that the ecosystem must be considered at its own level of organization. It was this perception that foreshadowed the current interest in hierarchies and led him to reject the terms "biotic community" and "complex organism" in favor of "ecosystem."

This chapter will begin constructing a concept of ecosystems. We approach the problem by analyzing how one must conceptualize any complex system, beginning with the nature of systems and the nature of complexity.

37

THE CONCEPT OF SYSTEM

The systems concept is simply a way to explain a part of the universe selected for study. A system consists of two or more components that interact (Hall and Fagan 1956), and it is surrounded by an environment with which it may or may not interact. Such a general definition emphasizes that systems represent an everyday component of human experience, not just in ecology but in the world of machines and social institutions as well.

The delineation of a system is somewhat arbitrary because a system is a construct of the human mind. Usually, however, there is some logic for deciding what to include and what to exclude because the components must interact in significant ways to produce the desired or observed behavior. These system components may be tangible, such as planets in a solar system, gears in a machine, or individual organisms in a population. They may also be abstract, like words in a language or ideas in a philosophical system.

In defining physical and biological systems, one is trying to gain an understanding of phenomena in nature. Systems are defined so as to leave our irrelevancies and simplify the number of components and/or interactions. As an example of the omission of irrelevant properties, one can safely ignore the color of a pendulum in measuring its frequency of oscillation. One may or may not be able to ignore other factors such as vibrations induced by road traffic. If other factors cannot be ignored, what one must define as the system becomes larger. As we will see later, putting bounds on a system is often easy in mechanical systems, somewhat harder in biological systems, and very hard in ecological systems.

Subfields of science generally focus on particular types of systems chosen to elucidate specific parts of the world. Physicists, for example, deal with mechanical, optical, thermodynamic, atomic, and many other types of systems for given purposes. Some classes of systems are seen as having a common underlying organization appropriate for describing diverse phenomena. For example, cybernetic organization seems appropriate for describing phenomena ranging from organisms and machines to corporations and societies. The task of choosing an appropriate system for investigating a particular phenomenon is inseparable from consideration of underlying organization and complexity.

It should now be clear why the concept of system is important in our present discussion. The system is a way of dealing with complex phenomena. It is a concept that is very natural to human thought in general and to scientific inquiry in particular.

Focusing more specifically on biological problems, Weiss (1971) defined a system as "a complex unit in space and time so constituted that its component subunits, by 'systematic' cooperation, preserve its internal configuration of structure and behavior and tend to restore it after non-destructive disturbances." His use of the words "cooperation" and "preserve" are unfortunate. There seems to be an implication that the components have a common aim and know how to achieve it. However, teleology is not implied in the definition, one simply observes the interactions as the causal explanatory principles of the system's behavior. Thus, the observation that phenomena occur at the ecosystem level and that we can consider the ecosystem as a system for the purpose of studying those phenomena does not imply any a priori commitment to the ecosystem as a supraorganism. The ecosystem is still a system even if it is conceptualized as having a quite different organization than the organism (Fredericks 1958).

It is precisely because of the complexity of ecosystems that an adequate concept of system is required. In many observation sets, the ecosystem is composed of a large number of parts and these parts interact in complex ways. For systems of this nature, it is possible to elucidate the behavior of the parts, but it is no trivial matter to arrive at the properties of the whole (Simon 1962). And yet, complexity of organization seems to be a key ingredient in ecological persistence in the face of external disturbances (Conrad 1976).

Major (1969) stressed the antiquity and universality of a concept of environmental systems. He noted this perception in diverse cultures as reflected in their language. Thus, we find terms such as tugai, tundra, steppe, chaparral, veldt, pampas, mallee, tiaga, and so forth, which indicate the recognition of local environmental systems. Major also emphasized that this perception of system has been formalized in a number of independent scientific traditions. Although ecosystem is the currently accepted term, there are many comparable designations: biogeocoenose, biocoenose, holocoen, biochore, epigen, nature complex, et cetera (see also Whittaker 1962). Thus, the ecosystem represents a perception of systems in the natural world that may be inherent in human experience.

This perspective was a key ingredient in many of the early geographical studies of natural history. It was certainly important in the development of plant geography (Chapter 2). Such a perspective was also central to the biotic community concept that grew out of biogeography. A strong systems perspective was implicit, for example, in the work of E. Forbes (1844) on the marine fauna of the Aegean Sea and in his concept of "provinces of depth." It is clear in Mobius's (1877) classic discussion of an oyster bank as a biocoenose or social community. It is present in S. A. Forbes's (1887) essay on the lake as a microcosm. It shows up again in the work of Warming (1909) on plant communities. The environmental system, in the form of a community, became a central focus of plant ecology in this century.

A system concept is implied in many early discussions of

environmental interactions. Geisler (1926), Jenny (1930), and Major (1951) discussed soil and vegetation as interacting components of a common system. McColloch and Hayes (1922), Taylor (1935a), and Jacot (1936) clearly saw soil and soil organisms as parts of a single system. Shelford (1931) stressed that plant and animal components could not be considered in isolation. W. S. Cooper (1926) argued that "organisms and environment make a system." Taylor (1936) discussed plants, animals, and their environments as parts of the "great unit system of matter and energy."

Thus, the concept of an environmental system is well established. However, the natural world is complex and can be viewed across a wide range of spatiotemporal scales. Unlike most mechanical systems, boundaries cannot be simply delineated and components cannot be easily designated. For this reason we must go beyond the concept of system and consider complexity.

TYPES OF COMPLEXITY

If complexity is the challenge faced by ecologists, we must first define what we mean by complexity. Complexity certainly increases as the number of components increases or, correspondingly, as the number of variables and parameters in a mathematical model describing the system increases. But a number of paradoxes are revealed by closer consideration of specific systems. First, there is the well-known example of the perfect gas. A typical macroscopic quantity of gas is of the order of 10^{23} molecules. Yet physicists are able to make thermodynamic predictions relating pressure, volume, and temperature not in spite of, but because of, the large number of atoms. At the opposite extreme, no exact prediction, in closed form, can be made concerning the motions of three bodies interacting through gravitational forces. Although the individual motions of the

41

10^{23} molecules are unknowable, this does not make any difference in thermodynamic predictions that are averages over the motions of all the molecules. When such averaging is impossible then problems may be insoluble. The three-body problem illustrates that the threshold at which a system becomes extremely complex may be very low.

Thus, complexity must be categorized by more than just the number of components. Weinberg (1975; Weinberg and Weinberg 1979) provides a classification of methodological approaches to systems of differing complexity. His classification follows those presented earlier by Weaver (1948) and Mandelbrot (Stent 1978).

Weinberg's first approach is typified by Newton. Newton made two critical assumptions that allowed him to reduce celestial mechanics to manageable proportions. First, he ignored all celestial bodies except the sun and planets. This greatly reduced the number of components he had to deal with. Second, he considered only pairwise interactions. Because of the mass of the sun, he could consider a set of independent subsystems, composed of the sun and each planet in turn.

It is clear that Newtonian approaches will only be successful in specific types of systems. They must have a small number of components with simple interactions. Weinberg used the term "organized simplicity" to characterize the systems that fall into this category. These "small-number systems" ordinarily yield to mathematical analysis, with each component described by a separate equation.

Whenever ecological observations can be explained in terms of a small-number system, this approach is to be recommended. Formal theories of population and community dynamics are based on this approach. The basic equations are similar to Newton's equations of motion. They include single-component dynamics (e.g., growth) and pairwise interactions (e.g., competition). Unfortunately, it is difficult to formulate some ecosystem problems in terms of a small-number system so that each component can be represented by a single equation.

The second approach reviewed by Weinberg (1975) is typified by the statistical mechanics approach to gases. In such "large-number systems," the components are very large in number, independent, and essentially identical. Component interactions are random, and overall system averages are easily performed. Weaver (1948) used the term "disorganized complexity" to characterize such systems, while Mandelbrot (Stent 1978) assigned them to the first stage of indeterminacy. In these systems, Newton's approach is impractical, and physicists therefore developed the formalism of statistical mechanics.

This approach has been less influential in ecology than Newtonian methods, although there have been some attempts to elucidate biotic structure using statistical approaches. For example, Kerner (1957, 1959) proposed a statistical mechanics theory for population models. There are also cases in which an overall average property can be derived. In decomposition studies, individual microbial populations are seldom followed. Instead a measure of the collective behavior of the total assemblage, such as respiration rate, is used. In many respects the search for singular, overall measures for a community, such as species diversity, reflects a similar orientation (Lane et al. 1975). However, in general, ecological observation sets do not involve very large numbers (e.g., 10^{23}) of "nearly identical" components as required by the statistical approach. The ecosystem also shows considerable internal structuring and is far from random. Therefore, the statistical mechanics approach is limited as a general paradigm for ecosystem theory.

The final class, "medium-number systems," includes most systems. These systems are characterized by intermediate numbers of components and structured interrelationships among these components. Examples of systems simple in number of components but complex in organization are nonlinear population models that exhibit chaotic behavior (May 1974; May and Oster 1976). Weinberg (1975) characterized these with Weaver's (1948) term: "organized complexity." Mandelbrot (Stent 1978) assigned these systems to the second stage of indeterminacy. Neither the mechanical nor the statistical approach will suffice for medium-number systems. There are too many components to describe each by a single equation; there are not enough to simply average properties.

Many ecosystem observation sets will require dealing with a medium-number system. As a result, mechanical and statistical methods adopted from the physical sciences are unlikely to be helpful for more than limited problem areas. The type of complexity found in ecosystem observation sets helps to explain why as much confusion as enlightenment could be introduced into ecology if mathematical methods are adopted simply because they are successful in physics and engineering.

Thus, the problem of the ecosystem scientist is not that the object of study is complex. Other complicated systems have yielded to scientific investigation. The real problem is the *type* of complexity involved in many observation sets. In general, ecosystems must be conceived as medium-number systems that can be expected to be particularly recalcitrant to analysis. The route to analysis is hinted at in the term "organized complexity." Any success in analyzing ecosystems is likely to be found in careful and explicit consideration of the organization underlying the complexity. And it is to this subject that we turn next.

THE CONCEPT OF ORGANIZATION AND THE CYBERNETIC APPROACH

What is "organization"? Clearly it is a property relating things in a system, but we must be more specific. Denbigh

(1975) has contrasted orderliness with organization. Orderliness is a measure of how well each specimen approximates an ideal. Crystals show orderliness. Organization is harder to define, but the very word, in suggesting "organic," leads one in the right direction. Denbigh classes all things that have organization as either "(1) organisms themselves, or (2) societies of organisms, or (3) nonliving structures or procedures created by organisms."

In addition to the association of organization with living things, several other properties seem pertinent. (1) An organization is something that exists independently of specific components. Individual trees may die, but the forest's organization remains. (2) The components of an organization are interdependent. A social insect, removed from its colony, seldom survives for long. (3) An organization seems to have a function. This is evident from the fact that most organizations we can think of consist of organisms and have functions in terms of fitness. (4) In the concept of organization, something dynamic, either past or present, is implied. (5) There is a sliding scale of organization. Two populations may simply coexist in an area, or they may be involved in a complex symbiotic relationship.

Cybernetic Organization

Just as we can classify systems as mechanical, chemical, social, and so on, various types of organization may be distinguished. For example, there are cybernetic organizations, in which information and feedback are emphasized. This type of organization has often been proposed for the ecosystem. It is clear that both transfers of information and negative feedbacks can be identified in ecosystem organization, and that insights have been gained by pursuing this approach (Patten and Odum 1981).

A cybernetic organization implies that the system is self-controlling. Through the establishment of feedbacks in-

45

volving the exchange of information as well as energy–matter transfers, system processes are maintained and controlled. Just as a mammal is able to maintain a constant internal temperature, the ecosystem maintains relatively constant rates of processing despite changes in the environment. If perturbed, the ecosystem is either homeostatic (i.e., returns to a constant equilibrium) or homeorhetic (i.e., returns to its preperturbation trajectory or rate of change). Since homeorhetic implies return to normal behavior and carries no implication of equilibrium, it seems a more appropriate term for ecological applications.

The cybernetic approach traces back to the concept of the community as a supraorganism. The approach was reinforced by Lindeman (1942) and his concept of trophic dynamics. Studies of the flow of energy through systems (e.g., H. T. Odum 1957) followed Lindeman's paradigm and led to a concept of ecosystems as carefully controlling their own functions.

The cybernetic approach per se began by combining a functional view of ecosystems as processors of energy, with information theory measures for species diversity (Patten 1959). The approach took advantage of the relationship between the information measures and entropy. The combination was used to construct an approach to ecosystems as self-controlled entities depending on a network of information exchanges and negative feedbacks.

Margalef (1963, 1968) searched for unifying principles by combining a cybernetic viewpoint with succession theory. He linked diversity indices (e.g., biochemical diversity) with measures of ecosystem metabolism (e.g., production/biomass ratio) in a model of successional changes from an immature to a mature state. Ecosystems were compared on the basis of their relative maturity.

Bosserman (1979) provided a detailed analysis of the properties that must be demonstrated for a specific ecosystem to qualify as a cybernetic system (i.e., "a self-organizing aggregate"). His analysis will serve to show both the advantages and limitations of the cybernetic approach.

One class of properties shows that self-organizing aggregates are integrated wholes. Such aggregates can be usefully considered as intact systems with macroscopic properties (e.g., nutrient cycling) that vary in a regular manner in response to energy and matter inputs. The system modifies its internal medium in a manner that permits continued existence. Such cybernetic properties capture an essential feature of ecosystems (i.e., the way that an ecosystem at any level of resolution is internally organized through the interaction of its components).

A second class of properties shows that self-organizing aggregates can be dealt with as legitimate scientific objects. They have definable functional and structural boundaries. More importantly, the macroscopic behavior of the system can be reproduced in space and time under operationally defined conditions.

The third class of properties, however, reveals the scale-dependence of the cybernetic approach. The self-organizing aggregate is seen as stable in space and time, in the sense of remaining within a small bounded range. The system is held together by cohesive forces that maintain it, along with its characteristic processes and subsystems, over a nontrivial but bounded life span. This third class of properties emphasizes the homeorhetic properties of the ecosystem and deemphasizes potential unstable behavior (i.e., the system moves to a new operating state or collapses completely). Thus, the cybernetic paradigm is appropriate in dealing with observation sets within which the ecosystem responds stably (i.e., homeorhetically) to perturbations. If the ecological problem requires us to deal with the natural world as distributed across a range of time–space scales, the paradigm is more limited. It is importantly limited in its ability to deal with situations in which the ecosystem responds unstably.

LIMITATIONS OF THE CYBERNETIC APPROACH

Ecosystems appear cybernetic in many observation sets. Therefore, we can expect the cybernetic approach to continue to be useful. Other important observation sets exist, however, within which the ecosystem is neither constant nor homeorhetic. Therefore, it will be necessary to examine critically the cybernetic concept and discover its limitations. Knowing these limitations, it may then be possible to expand the concept to deal with the full range of observation sets that must be considered in ecosystem analysis.

Problems with Equilibrium

The cybernetic approach is limited by its emphasis on equilibrium (or homeorhesis). Connell and Sousa (1983) reviewed a wide literature and showed the difficulties with demonstrating equilibrium in any ecological system. In general, a demonstration of equilibrium requires a longer data record than is available. In fact, the equilibrium concept derives from the "balance of nature" paradigm (see Chapter 2) and is useful only at small spatial scales viewed over short time intervals.

The difficulty in demonstrating equilibrium is illustrated nicely in a series of papers dealing with island colonization. Heatwole and Levins (1972) analyzed data on the reinvasion of defaunated mangrove islands by arthropods. They tried to demonstrate that the trophic structure of the arthropod community was the same before and after disturbance, even though the taxa occupying the trophic levels were different. Similarly, the trophic structure was more similar between islands than the taxonomy. Simberloff (1976), who collected the original data, protested that the data were not adequate to demonstrate that the trophic structure was at equilibrium. In particular, he argued that pooling arthropods from many small islands made the data look better than it was. It was stretching the data to conclude that the system was in equilibrium.

This, of course, is not to say that there is not a discernible pattern in trophic structure. In another study, Heatwole and Levins (1973) examined data from a single larger island adjacent to Puerto Rico. This study seems relatively free from the defects that Simberloff pointed out and yet showed that a trophic structure is established irrespective of the taxa involved. The argument is backed up by other studies such as those of Dammerman (1948) on Krakatau and Heatwole (1971) on the Coral Sea Islands. These studies establish the reasonable conclusion that detritus feeders and omnivores must be established before predators can successfully invade. Thus, the system appears to be cybernetic (i.e., self-organizing and self-controlling), but this can only be established within certain observation sets and at particular scales of resolution. That ecosystems are necessarily self-controlled, irrespective of scale, simply cannot be demonstrated.

A similar problem arises even if the criterion of equilibrium is expanded to include something like the shifting mosaic concept of Bormann and Likens (1979a). These authors considered (see Chapter 2) that a landscape may be constantly changing and yet an equilibrium may exist in the proportion of the landscape in each successional stage. This type of equilibrium has been confirmed by P. S. White (1979), by Zackrisson (1977) in northern Swedish boreal forests, and by Sprugel (1976; Sprugel and Bormann 1981) for balsam fir forests in the northeastern United States. And yet, when seeking a shifting mosaic equilibrium in Yellowstone National Park, Romme (1982) was forced to conclude that it did not exist, at least within the scale examined

(73 km²). Thus, it appears that even the shifting mosaic concept is scale-dependent. The equilibrium can be found at some scales and in some observation sets and not in others. It is possible to adopt an even looser concept of equilibrium. The system could be called cybernetic if it returns not to a unique equilibrium but to a preperturbation trajectory (i.e., homeorhesis). Alternatively, the system might be cybernetic if it remains within well-defined bounds. Unfortunately, these concepts do not avoid the critical problems. It would probably be more difficult to demonstrate these behaviors with field data than it is to demonstrate equilibrium. These concepts are still limited since there are some observation sets in which ecosystems respond unstably and therefore do not return to a prior trajectory or remain within reasonable bounds.

Problems with Instability

The cybernetic approach is problematic in addressing practical questions that involve ecosystem instability. Because the approach emphasizes the ways in which the system is regulated under normal conditions, it has been less useful when applied to problems in which ecosystems are stressed beyond the limits of the self-regulatory mechanisms. In these cases the system responds unstably and the homeorhetic analogy breaks down. Thus, there appears to be a significant class of problems for which the cybernetic viewpoint is incomplete.

Viewing ecosystems as cybernetic tends to emphasize control mechanisms. Though both positive and negative feedbacks can be found in cybernetic systems, the cybernetic viewpoint has focused on negative feedbacks operating in ecosystems. A number of negative feedback relationships can be extracted from our knowledge of ecological systems. For example, as a consumer feeds more rapidly, less food remains, thereby decreasing the feeding rate. Indeed, it can be reasonably maintained that such negative feedback relationships are important or even crucial in maintaining the normal operating state of the system. However, not everything about ecosystem behavior is well regulated. Many population phenomena are more easily conceptualized as unstable, positive feedback systems. Populations tend to grow exponentially until they meet some constraint imposed upon them from outside. Such phenomena appear less as balanced self-regulation and more as runaway processes constrained by a barrier. In fact, the ubiquity of positive feedback processes in ecosystems is now becoming more apparent (DeAngelis et al. 1986). Therefore, insofar as the cybernetic approach tends to emphasize stable, negative feedback mechanisms to the exclusion of unstable, positive feedbacks, the approach must be seen as an incomplete paradigm for ecosystems.

Problems with Engineering Mathematics

In many respects, cybernetics is not so much an organizational principle forced upon us by the natural world as an approach borrowed from engineering. The logic involved is straightforward: within some observation sets, ecosystems appear as well-regulated systems. Such cybernetic systems are well known in mechanical and electrical devices and their principles have been studied in detail. Therefore, it is logical to attempt to explain the observed phenomena using these same engineering principles.

The assimilation of engineering approaches has had the salutary effect of introducing the ecosystem scientist to mathematics. The power and rigor of mathematics is an important adjunct to any scientific field. However, "systems" modeling, imported from engineering, carried with it considerable baggage. In its original context, the purpose of systems analysis was to combine parts so that the resultant system performed according to specifications. Thus, rather than fostering a perception of whole environmental systems, such approaches tend to focus attention back on preconceived homeostatic or homeorhetic behavior.

Waide and Webster (1976) suggested that design constraints that might have structured ecosystems are not well understood. Therefore, an array of techniques from systems engineering may be less appropriate to ecosystems than to some other applications. Hill and Durham (1978) similarly outlined a variety of assumptions implicit in control theory which appear not to apply in ecology. Rosen (1972), Pattee (1972), and Simon (1973) detailed other objections to the use of classical systems approaches for studying biological systems.

T.F.H. Allen and Starr (1982) referred to large-scale ecosystem simulation models that follow the cybernetic paradigm as brute-force reductionism. While this indictment seems overly harsh, model construction often begins with an arbitrary definition of components and their interactions. Behavior of the total system is then calculated as the result of the component behaviors. The approach does not begin by considering ecosystem phenomena at their own spatiotemporal scale. It is probably true to say that early developments in the subdiscipline of systems ecology did much to entrench the cybernetic view of ecosystems.

Problems with the Cybernetic Approach

Seen in the preceding context, discussions of whether ecosystems are (Patten and Odum 1981) or are not (Engelberg and Boyarsky 1979) cybernetic are somewhat off the point. It is clear that there are some spatiotemporal frameworks and some observation sets where ecosystems do not appear homeostatic or homeorhetic. Therefore, to maintain that the cybernetic approach is the only or the most fundamental way of conceptualizing ecosystems will lead to more controversy than insight.

Two important points need to be extracted from our analysis of the limitations of the cybernetic approach. First, as with other points of view we reviewed earlier, to say that the cybernetic view is limited should not be taken to imply that it is useless. Studies of the control mechanisms operating in ecosystems and mathematical modeling based on feedback mechanisms will continue to advance our understanding and must be incorporated into any adequate theory. However, the paradigm is limited and only refers to some possible behaviors of ecosystems.

The second important point is that designating ecosystems as systems does not necessarily imply that the systems are cybernetic. A system with a strictly cybernetic organization that operates on a single time–space scale is only one of the possibilities. Therefore, to consider ecosystems as medium-number systems and to seek the principles that underlie their organized complexity does not imply any adherence to a supraorganism or equilibrium point of view.

CONCLUSIONS

To develop an adequate theory, it appears useful to conceptualize ecosystems as complex systems. Our analysis of complexity indicates that ecosystems show "organized complexity." Therefore, the route to pursue in understanding ecosystems will be to analyze organization. However, we have found that the assumption of a cybernetic organization cannot deal with all of the observation sets of interest in ecosystem analysis. Cybernetics is neither the fundamental nor sole organizing principle of ecosystems.

While the cybernetic approach has been and should continue to be useful, it is limited. Some observation sets are usefully explained in terms of a system that organizes and controls itself. When one is dealing with such an observation set, the cybernetic paradigm is valuable. But this is a far cry from saying that ecosystems are cybernetic. This implies that there is some scale-independent "thing" out "there" that is self-controlled. It is certainly not true that ecosystems viewed at any of the possible scales of observation will always be self-controlled. The paradigm has little to offer if a problem requires prediction of what will happen when a system responds unstably. Therefore, cybernetics cannot be considered as a complete theory of ecosystems, capable of explaining all of the relevant phenomena. Self-organization and self-control must be incorporated into any adequate theory, but this must be done within a broader conceptual framework. Collier et al. (1973) maintained that homeostasis is a characteristic of all ecological systems. But they see this process as scale-dependent, quite different at different hierarchical levels.

In the next chapter, we will propose that a more fundamental organizing principle can be found in hierarchy theory. The remainder of the book will be occupied with detailing what is implied in this proposal and what theoretical and empirical evidence can be called upon to show that ecosystems can be usefully conceptualized as hierarchical systems.

CHAPTER 4

The Concept of Hierarchy and Its Typical Application

Many authors have pointed out the relevance of hierarchy theory to the study of complex systems. Mesarovic and Macko (1969) suggested that the theory resolves the dilemma, inherent in the study of complex systems, between simplifying details and accounting for observable behaviors. Simon (1962) proposed that hierarchy theory can decompose medium-number systems by focusing on an internal redundancy of structure. Later he argued that hierarchical organization is associated with a fundamental parsimony in the interactions among components (Simon 1973). Thus, hierarchical descriptions help to manage complexity by isolating dynamics at a single level and ignoring details at lower organizational levels.

The concept of hierarchy has a long history in science (Whyte 1969) including biology (Novikoff 1945; Leake 1969). It has been formalized into a mathematical theory (e.g., Mesarovic et al. 1970) and has been used to examine methods of explanation in science (Feibleman 1954). Hierarchy theory is not new to ecology either (Schultz 1967, 1969). As early as 1942, Egler recognized that organism, community, vegetation, and ecosystem formed an arbitrary hierarchy that was useful in looking at ecological problems. A similar understanding can be found in the writings of Bray (1958) and S. Wright (1959) who discussed the ways in which genetics operates at different levels in the hierarchy. Rowe (1961) recognized that any object of study in ecology must be composed of lower-level subsystems and must itself be part of levels above.

Major credit for the introduction of hierarchy theory to ecosystem ecology must be given to Overton (1972, 1975) who outlined the advantages of this approach over a decade ago. Based on the general systems theory of Klir (1969), Overton stressed the importance of decomposing complex ecosystem dynamics into simpler components (Overton 1974). Component models could then be developed and reassembled using the computer processor developed by C. White and Overton (1974).

Applications of hierarchy theory can be found in a number of theoretical studies, including Reichle et al. (1975), Patten et al. (1976), and T.F.H. Allen et al. (1977). McIntire and his colleagues (1975, 1978) used the concepts of Overton (1972, 1975) to develop models for stream ecosystems. Shilov (1981) discussed the applicability of the theory to a number of problems. O'Neill and Waide (1981) applied the theory to toxic effects in ecosystems. T.F.H. Allen and Iltis (1980) and Allen and Starr (1982) provided an extensive review of the basic concepts and their application to ecology.

Pattee (1978) argued that we normally adopt hierarchical explanations when dealing with complex systems. We look to higher levels for significance and to lower levels for mechanisms (Rowe 1961; Overton 1977; Webster 1979). Collier et al. (1973) saw that mechanistic explanations of ecosystem function were to be sought through the study of processes involving ecosystem components. At the same time, they recognized that the significance of ecosystem processes was to be sought at the level of major regional ecosystem types or perhaps the biosphere as a whole. Pianka (1974) and MacMahon et al. (1978) discussed the organism as participating at many levels of ecological organization. To understand the significance of some organismal behavior requires recognition of higher levels in which the organism participates. Thus, to understand the significance of re-production requires that we consider the organism as a component of the population or species.

Such studies suggest that hierarchy theory is relevant to ecosystem analysis. However, these studies have not taken full advantage of the theory. Many of the applications are based on a simple view of hierarchies (Guttman 1976; Bossert et al. 1977) or on a limited concept of the ecosystem. In other cases, the discussion focuses on human perceptions of tangible and bounded objects (Rowe 1961; MacMahon et al. 1978; Webster 1979). Any of these modifications can change key components of the theory and rob it of much of its power to deal with the organizational relationships of ecosystems.

On first view, then, hierarchy theory recommends itself as useful for analysis of complex systems and as applicable to ecosystem studies. However, hierarchy is a commonly used concept in biology and we must proceed carefully and systematically. First, we must examine the concept of a hierarchy.

THE CONCEPT OF A HIERARCHY

Consider the following control problem. A mobile solar collector can run back and forth on a stretch of tracks. The tracks run below trees, so this pathway is subjected to spatially and temporally varying patterns of light and shade. The problem involves automating the vehicle so the collector makes optimal use of available sunlight.

If the vehicle moves steadily back and forth, it will avoid being stuck in the shade, but this solution is far from ideal. A better idea is to equip the motor with a light detector. When the detector indicates that solar radiation is greater than some value, R_1, the vehicle stops. It does not move again until the light level at that spot falls below R_1. However, the device will stay parked in places whose solar radiation levels are below what is available nearby. To alleviate this disadvantage, one might add an instrument to detect gradients in the solar radiation. If the gradient dR/ds (s is distance) is positive in the current direction of motion, the vehicle continues to move.

However, the collector is still liable to get stuck at a local radiation maximum, R_{max}, which is low compared to what is available at some distant part of the track. A solution to this problem is to install a still higher control that overrides the lower control systems and triggers renewed searching whenever R_{max} falls below some level. One can imagine a still higher level of control that causes the search to shut down temporarily when no profitably high levels of radiation can be found anywhere along the track.

What we have described is a control hierarchy. The purpose of the hierarchical system is to collect as much sunlight as possible. To an observer, the mechanical device appears to be displaying purposive behavior. Each level in the hierarchy can be overridden by the next higher level, and is thereby under the constraint or control of the next higher level. The higher-level control in a sense is pursuing a more general strategy to which the more local strategy of the lower-level controls are subordinated.

Each level of the hierarchy corrects for errors or inefficiencies in the levels below. This observation calls to mind the parable of the watchmakers that Simon (1962) used to illustrate hierarchical structures. According to this parable, Chronos and Tempus assemble watches with 1,000 parts. Any interruption causes the current assemblage to fall apart. Tempus puts the pieces together sequentially and an interruption causes all preceding work on that watch to be lost. Chronos puts together stable subassemblies of 10 pieces each. An interruption at piece 608, for example, disrupts only the last subassembly of eight pieces. The other 60 subassemblies are not harmed. The point of the parable is that Chronos used a two-level hierarchical procedure so that a failure caused by an environmental perturbation is localized. We shall see in Chapter 6 that all biological organizations can be conceived in terms of stable or metastable subunits bound into larger units.

Hierarchies in Biological Systems

Obvious cases of hierarchies can be found in the activity of any organism. When a cat springs on a mouse, for example, a decision is first made in the cat's brain. Nerve impulses are then sent to several subsystems, the various muscles involved in the spring. Each muscle is itself a structural hierarchy. It consists of bundles of muscle cells, each of which is composed of many fibrils, which in turn are made up of many filaments of interdigitating mysin and actin macromolecules. When the nerve impulses reach the nerve–muscle junctions, signals are released that, through the intracellular intermediaries of Ca^{2+} and ATPase, cause the actin and myosin macromolecules to link together into actomyosin. This causes a shortening of each filament. These shortenings are coordinated into overall contractions of the muscle, resulting in the cat's spring.

Numerous other examples of hierarchical action in biological systems could be mentioned; for example, the action of operator and structural genes in subcellular processes. Each of these would be far more complex than our example of the solar collector. The control hierarchy of the solar collector was relatively straightforward, with only one subsystem at each level. Usually, a given hierarchical level has several subsystems, such as the many muscles in a cat. In many interesting biological hierarchies, however, the number of subsystems on a given level may be relatively few. For example, as Simon (1969) pointed out, only twenty amino

59

acids are involved in all the diverse proteins of the next hierarchical level.

In studying ecosystems we implicitly use many hierarchical assumptions. One of these assumptions is the relative independence of subsystems on a given level. An example is the earth's total biosphere, which can be divided into many distinct ecological subsystems or biomes. The tundra and tropical savanna biomes are relatively independent and usually can be studied without reference to each other.

Goodall (1974) suggested that we make more conscious use of hierarchical thinking and take advantage of the decoupling that often occurs between subsystems on the same level. He considered a grazing system in which cattle and wallabies feed on grass. Rather than dealing with all the complexities of this system, a hierarchical structuring might help. Thus, two distinct conditions for the grass exist: (1) that of being in the open and (2) that of being under the canopy. These are relatively independent of each other and make convenient subsystems. Under the canopy there are two main grasses, *Themeda* and *Heteropogon*, which may in turn be thought of as subsystems within the "undercanopy" system. This procedure can be carried out to whatever degree of detail is desired.

The above description of hierarchies has been informal and anecdotal. Most biologists are already familiar with the general idea of hierarchy through the concept of "levels of organization" (cell, organism, population, community). However, even our informal discussion should make it clear that the term hierarchy is not restricted to this simple sense.

In fact, the familiarity of the biologist with levels of organization represents a real impediment to the presentation of our concept of ecosystem organization. Ecologists already use hierarchy theory when they divide ecosystems into trophic levels and trophic levels into populations. Therefore, before we can proceed to a more formal and systematic coverage of hierarchy theory, it will be necessary for us to point out the shortcomings of some of the possible applications of hierarchy to ecosystems.

THE SIMPLE ECOLOGICAL HIERARCHY

In ecological literature, hierarchy is usually identified with the concept of levels of organization. This concept arrays biological systems into higher and higher levels of organization. In the simplest series (cell, organism, population, community, ecosystem), each level is composed of the subsystems on the next lower level and is controlled by the level above it. It is clear that ecological organizations show hierarchical structure, but it should also be clear that the simple series is unlikely to be useful across the range of observation sets and spatiotemporal scales involved in ecosystem analysis.

In Chapter 1 we saw that the population–community viewpoint was inappropriate for explaining some ecosystem phenomena, such as nutrient cycling. Nevertheless, a naive application of levels of organization might lead one to believe that ecosystems are composed solely of biotic entities such as communities and populations. Therefore, all ecosystem processes, including nutrient cycling and energy flow, should be reducible to populations and their interactions.

Problems with Simple Reductionism

Man is always reductionist in his mechanistic explanations. Indeed, when we say that we have explained a phenomenon we ordinarily mean that we have shown the phenomenon to be the consequence of interactions among system components (Overton 1977; Webster 1979). Therefore, it is not surprising that ecologists have always subdivided environmental systems to explain ecosystem phe-

61

nomena. The problem is that too simple a hierarchical concept can lead one to assume that the only way to subdivide the ecosystem is into species populations. Ecosystems certainly have species populations in them, and for many problems populations are the relevant components. But it does not follow that species are always the most useful subunits to explain ecosystem phenomena (Webster 1979).

Perhaps the first challenge to naive reductionism occurred in the nineteenth century with the attempt to explain the behavior of gases and liquids with the classical dynamics of individual molecules. A rigorous description proved to be impossible and physicists were led to classify laws into different categories. Eddington (1939) distinguished between primary laws that govern individual particles and secondary laws, such as the second law of thermodynamics, that govern large aggregates of particles. Failure of strict reductionism in the most reductionist of sciences, physics, suggests that naive reductionism will also fail in ecology.

Guttman (1976) points out the assumptions involved in naive applications of hierarchy. First, it is assumed that every system at level n is made entirely and exclusively of level $n-1$ subsystems. This assumption is clearly violated for ecosystems composed not only of populations (i.e., $n-1$ subsystems) but also of abiotic components. Second, it is assumed that to explain level n behavior, all you need to study are isolated interactions at the $n-1$ level. Once again this assumption is violated, for ecosystem dynamics involve interactions between biotic and abiotic components as well as interactions among populations (Webster 1979).

The real problem with the simple hierarchical concept is that it insists on one and only one class of level $n-1$ subsystems. With some exceptions (see chapter by C. J. Walters in Odum 1971), ecology textbooks conceive ecosystems as built up from their constituent communities. The discussions may leave the impression that no other subsystems at the $n-1$ level are useful for explaining ecosystem dynamics. The presentations seem to have lost the earlier insight that it was observer perception that determined the relevant levels (Rowe 1961; MacMahon et al. 1978) and that the levels could change significantly for different problem areas (Egler 1942).

In spite of the problems, the use of populations to explain ecosystem phenomena seems very logical. Clearly, there can be no ecosystem without populations. Organisms formed into populations are tangible objects, easily identified and studied. Our evolutionary history has taught us to place special value on such tangible objects. In addition, when we focus on populations we bring to bear considerable developments deriving from the theory of natural selection, concern for zoogeographic distributions of organisms, and demographic theories of animal populations. This tradition has produced a powerful and consistent theory of population and community dynamics.

Since the explanation of ecosystem phenomena in terms of population interactions seems so logical, why should anyone object? An initial answer to this question has already been given in Chapter 1. There we pointed out the ambiguities in the ecosystem concept resulting from considering the population–community or the process–functional as the only or the most fundamental way to view the ecosystem. At that point we indicated that the population approach was particularly awkward for addressing phenomena such as decomposition. In these cases, populations and even organisms are hard to define and abiotic components of the system are just as important as biotic components in explaining phenomena (Webster 1979).

In fact, the existence of abiotic functional components is a fundamental problem with breaking ecosystem structure into subsystems exclusively composed of populations. Although ecology is often defined as the investigation of relationships between organisms and their environment, there is a remarkably fuzzy notion of just what the environment is. Ecology typically defines its organizational unit as somehow separate or distinct from its environment. This viewpoint destroys the fundamental unity of organic entities and their environments and has led to confusing attempts to systematize environment as separate from the organism (Haskell 1940; Mason and Langeheim 1957; Maelzer 1965). To be useful for the development of theory, an environment concept must recognize the fundamental unity between organisms and their environment (Patten 1978, Webster 1979).

We should not be too scandalized by the problems ecologists have with incorporating the environment as a functional component of their systems. The same problem exists throughout biology due to the naive application of hierarchical concepts. For example, it is certainly true that organisms contain cells and that it is impossible to have an organism without cells. However, bones and body fluids become something of an embarrassment when we maintain that organismic phenomena can be reduced to nothing but cells. In fact, it is as difficult to explain organismic physiology exclusively in terms of cellular metabolism as it is to explain ecosystem phenomena exclusively in terms of what we know about populations.

Perhaps the best illustration of the dangers inherent in the simple ecological hierarchy is the stability–diversity controversy. MacArthur (1955) originally posed the hypothesis in terms of the diversity of energy-flow pathways in ecosystems. However, later investigators focused almost entirely on species diversity. The result was an attempt to infer an ecosystem property, stability, using only information about species. As we will point out in Chapter 7, it has proven impossible to verify the relationship between stability and complexity when the observation set is restricted to species populations.

The Trophic-Level Concept

Historically, the trophic-level concept has been an important paradigm for subdividing ecosystems. Species populations are grouped into levels based on the number of trophic links between them and the primary energy source, the sun. The trophic-level concept has been well established since the 1920s, for example, in Elton's (1927) pyramid of numbers. Lindeman (1942) reformulated the concept in an ecosystem context by emphasizing the role of decomposer organisms. He also introduced a dynamic dimension by linking the trophic-level approach with succession theory. In so doing, he established trophic dynamics as a key component of emerging ecosystem theory, and it is tempting to think that trophic structure may form the basis for a hierarchical view of ecosystems. However, trophic levels are simply another attempt to reduce ecosystem dynamics to population dynamics.

Difficulties in unambiguously assigning species to trophic levels cast doubt on trophic dynamics as the organizing principle of ecosystem theory (Koslovsky 1968). Consider the food web (Fig. 4.1) constructed by S. M. Adams et al. (1983) from reservoir data. A given species or group of species might be the common resource of more than one consumer species. Some species are easy to assign to levels, but others are not. For example, a fish that feeds on phytoplankton only is assigned to level 1 and a fish that feeds on herbivorous zooplankton alone is assigned to level 2. But how would one define a fish that feeds on both? An assignment of trophic level can be made, but it depends on knowing the proportion of feeding the fish does on its two resources. Assume the fish's diet is 25 percent phytoplankton (level 0) and 75 percent zooplankton (level 1). Then we de-

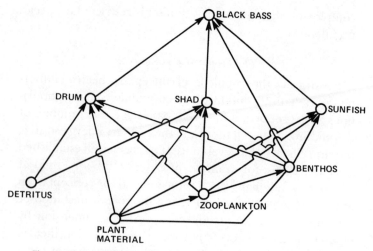

Fig. 4.1. Food web diagram from S. M. Adams et al. (1983). The complexity of trophic interactions makes it difficult to aggregate populations into discrete trophic levels.

fine, following Carney et al. (1981) and S. M. Adams et al. (1983), trophic level T_f as

$$T_f = 1.0 + 0.25(0) + 0.75(1) = 1.75.$$

In general we can assign a trophic position to any population by summing across resources. Therefore, a trophic position is defined, though it is no longer restricted to integer values.

Even the food web pictured in Figure 4.1 omits much of the complexity of the natural world. Ulanowicz (1983) has emphasized the importance of recycling through detritus and detritivores, as exemplified in the Cone Springs ecosystem (Fig. 4.2). The designation of the trophic position of species i (T_i) now becomes even more complex. We must consider L_{ij}, the number of steps that a given unit of energy takes to reach species i from species j, and P_{ij}, the probability that the given unit of energy has reached species i by pre-

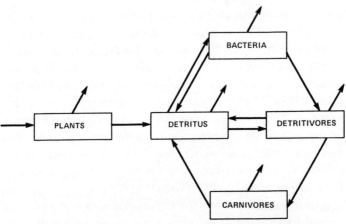

Fig. 4.2. Conceptualization of the Cone Springs ecosystem (Ulanowicz 1983). The recycling of material through detritus and detritivores makes it difficult to describe the dynamics of this system in terms of discrete trophic levels.

cisely this path. Then, T_i can be defined by the sum of the $L_{ij}P_{ij}$ over all resources j (DeAngelis 1980).

However, while it is feasible to assign species to trophic position, the levels themselves are no longer discrete entities. As a result, hypotheses about the behavior of whole trophic levels (Hairston et al. 1960; Slobodkin et al. 1967; Wiegert and Owen 1972) may be unfalsifiable due to difficulties in the trophic-level concept itself (Murdoch 1966; Ehrlich and Birch 1967). It is now well recognized that the trophic-level concept is most useful as a heuristic device and tends to obscure, rather than illuminate, organizational principles of ecosystems.

Hierarchies and the Problem of Scale

The problem of defining ecosystems in terms of populations is also inherent in seeking boundaries for the ecosystem based on distinct groupings of species populations along environmental gradients. The boundaries of an ecosystem (e.g., a watershed) do not necessarily correspond to a sharp change in species. The problem also arises when productivity of a lake is derived from physiological parameters of individual algal species (T.F.H. Allen and Starr 1982), and again when model ecosystems are built from interacting Lotka–Volterra equations for individual species (May 1973a; Maynard Smith 1974). The simple fact of the matter is that the population is a scientific object of interest at a specific scale of observation. To insist that all phenomena that occur at other scales of observation must be reduced to population dynamics is an improper application of the concept of hierarchies (Webster 1979).

Specifying the proper spatiotemporal scales on which to view the dynamics of the natural world requires ecologists to go beyond their own perceptual levels. They must conceptualize the system on the scale at which phenomena are actually observed. When the proper spatiotemporal scale is designated, the appropriate level of ecological organization can then be made clear. Then subsystems can be defined objectively as those needed to explain this specific level of observation (Overton 1977).

Levins (1973) perceived these problems of scale and argued that we cannot accept commonsense components as the best decomposition of the systems of interest. Rosen (1972) agreed, and suggested that biologists should have no preconceived notions about appropriate subunits. Subunit specification should be allowed to emerge from analysis of system dynamics. He observed that in physical and engineering sciences, fractionation is based on two far-reaching assumptions: (1) each subsystem can be studied in isolation, and (2) any property of the system can be reconstructed from relevant properties of the subsystems. These assumptions hold for broad classes of physical systems, because the components retain their properties when studied in isolation. However, it is not clear that such assumptions can be applied to ecosystems. Ecosystem components can behave

68

very differently when isolated from the larger system of which they are a part. Clearly, populations can achieve growth rates in the laboratory that are never realized in the field. The common problems of extrapolating laboratory measurements to the field should be sufficient demonstration that measurements on isolated components may provide little information about the behaviors of these components as functioning parts of ecosystems.

Rosen (1972) showed that physical fractionations of biological systems produced subsystems that were recognizable and thus had independent biological meaning. However, he was unable to reconstruct intact system dynamics on the basis of interactions among these subsystems. He then showed that he could fractionate system dynamics into subsystems from which the intact system's behavior could be reconstructed. However, the subsystems no longer corresponded to a human perceptual scale and bore no relationship to subsystems defined by human experience. The subsystems were the relevant components of the system's behavior but were not scientific objects in their own right.

Rosen later (1977b) returned to this analysis in the context of measurement theory. He pointed out that all of our studies of complex systems are done through the agency of measurements and measurement instruments. The designation of "natural" components of a system is often restricted to those components that we have measured in the past and for which we have developed instrumentation. Thus, natural components, like populations, are more often a circumstantial result of the way we have looked at the system in the past. There is no guarantee that such components will be adequate or convenient for explaining the system viewed in a new way.

Rosen (1977b) is particularly emphatic about the relative nature of any description of the system. Our description depends on a particular set of instruments, a particular way of looking at the system (i.e., a particular observation set). Because we do not have available to us all possible measuring instruments, we must never consider that any one description is absolute or necessary. The real problem with reductionism is not that it advocates analysis into simpler subsystems. This is necessary to arrive at mechanistic explanations. The problem is that reductionism advocates, in advance, one set of subsystems, which it posits to be the only admissible set of components.

The implications of Rosen's analysis for ecology are clear. Some ecologists have insisted that ecosystem components be recognizable scientific objects in their own right. Because interesting problems exist at the population and interpopulation level, they must necessarily be the relevant explanatory subunits of ecosystems. This argument insists that ecosystem subunits have a direct physical relation to man's perceptual experience. Such an approach can seriously limit ecological theory. Just as ecology is in need of a better concept of space–time scales, so it must expand its ability to recognize theoretically significant ecosystem components beyond the narrow bounds implied in the level-of-organization concept. What is needed is a "function-preserving" fractionation that is compatible with original system dynamics (Rosen 1972). Weiss (1971) perhaps stated the point most clearly: "The state of the whole must be known to understand the collective behavior of the parts."

SUMMARY

A number of authors have argued for the relevance of hierarchy theory to ecology in general and ecosystem analysis in particular. Indeed, through the levels-of-organization concept, most ecologists are already aware of the usefulness of simple hierarchies. However, as typically applied, the level-of-organization concept implies that ecosystem processes should be explained by exclusive recourse to community or population dynamics.

Our analysis indicates that there is a danger in such a simplistic application of the concept of hierarchy. Because ecologists are familiar with "levels of organization" does not guarantee that they are properly applying the hierarchy concept. Once again we are faced with problems of spatiotemporal scale and with limited views of the natural world. Because the population can be fruitfully studied at its own level of resolution does not mean that phenomena at other scales can be reduced to population interactions. Because some large-scale problems (e.g., some aspects of food web dynamics or community structure) can be fruitfully explained in terms of population interactions does not mean that all ecosystem phenomena must be explained in this way. The attempt to explain all ecosystem phenomena in terms of populations is simply another form of the unacceptable reductionism, which maintains that all biological phenomena can be explained in terms of physics and chemistry.

Such a conclusion unnecessarily restricts the explanatory tools available to the ecosystem analyst. Therefore, you should not assume that you are already familiar with the scope of hierarchy theory because you have used the level-of-organization concept. Hierarchy theory is more broadly conceived than any single series of levels. In fact, as we will begin to develop in the next chapter, the theory is most concerned with reconciling the broad range of spatiotemporal scales over which we observe complex ecosystems.

Part III

A Proposal for a Theory

The first four chapters were designed to question the current theoretical underpinnings of ecosystem analysis. In Part I (Chapters 1 and 2) we found that both biotic and functional approaches to the ecosystem were limited. A historical succession of problems caused certain spatiotemporal scales to be emphasized to the exclusion of others and left us heir to questionable static or equilibrium views of the natural world. In Part II (Chapters 3 and 4) we found that the concepts of system and hierarchy were useful but that current applications of these ideas did not encompass the full range of spatiotemporal scales needed to deal with the organized complexity of ecosystems.

The next two chapters will propose hierarchy theory as a theoretical framework for ecosystem analysis. Like any scientific theory, hierarchy theory will ultimately prove to be another limited approach. Thus, we will not argue that the natural world must be, in fact, hierarchically structured. To make use of the theory, it is sufficient that many observation sets permit ecological systems to be conceptualized as hierarchical. Our point is simply that the theory provides a means for dealing with organized complexity. Using the theory, the ecosystem can be dealt with as a small-number system. We will proceed by presenting the elements of the theory and then show that the hierarchy concept is both reasonable and productive in leading to new knowledge.

We are conscious of the dangers of proliferating new theoretical constructs. Ecology has many general theories but few have been fruitful in producing new insights. Therefore, we have taken seriously the challenge of demonstrating that the theory we propose is not simply an arbitrary imposition of yet another faddish idea. We propose that the following tests need to be applied to hierarchy theory or any other new construct:

(1) *The theory must be internally consistent.*
(2) *The theory must not be adopted simply because of success in other fields.*
(3) *The theory must agree with known properties of ecosystems.*
(4) *The theory must be capable of producing new and testable hypotheses.*

Chapter 5 reviews the elements of hierarchy theory and tries to show that the theory satisfies the first criterion. Since a significant literature has developed on hierarchy theory, both in general systems research and in ecology, we present only those aspects of immediate relevance to ecosystem analysis.

In Chapter 3 we suggested that the adoption of engineering mathematics into ecology might not pass the second criterion. In Chapter 6 we will argue that current concepts of the origin of life and early evolution make it logical to assume that biological systems are hierarchically structured. Indeed, hierarchical structuring may be one of the only feasible architectural plans for evolving a complex medium-number system.

The final two tests, criteria 3 and 4, must wait for Part IV of the book. There we will develop the theory further and show how it can be productive of new insights.

CHAPTER 5

Some Elements of Hierarchy Theory

Hierarchy theory has been developed primarily in the context of general systems theory. Comprehensive treatments can be found in Simon (1962, 1969), Whyte et al. (1969), and Pattee (1973). Bunge (1959a, b, 1969), Feibleman (1954), and Grene (1967) deal with the metaphysical and epistemological aspects of hierarchy. Mesarovic et al. (1970) apply the theory to control engineering. Mathematical and simulation tools for ecological applications are presented by Patten (1978, 1982) and Overton (1972). T.F.H. Allen and Starr (1982) provide a complete overview designed for an ecological audience. Since these references adequately introduce the general concepts, we will limit the present chapter to those elements of hierarchy theory immediately relevant to ecosystem theory.

STRUCTURE BASED ON DIFFERENCES IN RATES

Central to the theory is the concept that organization results from differences in process rates. This fundamental insight was introduced by Simon (1962, 1969, 1973). Much of this chapter is based on his work and on refinements presented by T.F.H. Allen and Starr (1982).

Medium-number systems, like ecosystems, operate over a wide spectrum of rates. Behaviors can be grouped into classes with similar rates, and if the classes are sufficiently distinct, then the system can be considered as hierarchical

75

and dealt with as a small-number system. The structure imposed by differences in rates is sufficient to decompose a complex system into organizational levels and into discrete components within each level (Overton 1974).

Vertical Structure in Hierarchical Systems

System behaviors must first be arrayed into levels of organization. Behaviors corresponding to higher levels occur at slow rates. Conversely, lower organizational levels exhibit rapid rates. For example, individual tree leaves respond rapidly to momentary changes in light intensity, CO_2 concentration, and the like. The growth of the tree responds more slowly and integrates these short-term changes. Change in the species composition of the forest occurs even more slowly, requiring decades or even centuries.

The resulting vertical structure may be either nested or nonnested. In a nested hierarchy, the higher levels are composed of and contain the lower levels. Thus, the forest is composed of the trees and the trees are composed of their individual parts. In a nonnested hierarchy, the higher level is a distinct entity and does not contain the lower levels. For example, in the pecking order of a flock of birds, higher-level birds control lower-level individuals but do not contain the lower levels in any sense. For the purposes of developing ecosystem theory, we will limit our discussion to nested hierarchies.

Levels in the hierarchy are isolated from each other because they operate at distinctly different rates. The overall organization is "nearly decomposable" (Simon 1973) because each level can be segregated on the basis of response times. It will profit us to develop this concept of behavioral isolation in more detail.

Consider the black box in Figure 5.1. This box is a linear input–output device, far simpler than a level in an ecological hierarchy, but it will serve nicely for illustration. As sig-

INPUT   **OUTPUT**

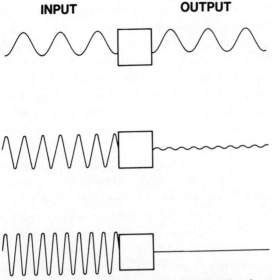

Fig. 5.1. Attenuation of an input signal by a linear "black box." Low-frequency signals (top) pass unmodified while higher frequency signals (middle and bottom) have their amplitudes decreased. The figure illustrates how a level in a hierarchy can be isolated from high-frequency (i.e., rapid) dynamics at lower levels in the hierarchy.

nals pass through the box, they can be attenuated or amplified and shifted out of phase (e.g., Child and Shugart 1972). Inputs at low frequencies will not be attenuated (top). The amplitude of the output will be equal to or greater than the amplitude of the input. At higher frequencies, characteristic of the device, the input signal begins to be attenuated (middle). The frequency in radians per unit time at which this attenuation begins is the reciprocal of the time constant characteristic of the box (see Waide and Webster 1976 for details). As the frequency of the input signal is increased, the attenuation also increases (bottom).

This example shows how successive levels in a hierarchy can be sealed off from each other. Each level acts like a filter, altering the signal it receives by attenuating frequency

77

components greater than its own characteristic frequency. Therefore, high-frequency dynamics are confined to lower organizational levels (Overton 1977). Signals passing upward in the hierarchy become dominated by fewer distinct components having reduced frequencies (T.F.H. Allen and Starr 1982). Levels that differ in modal frequency by as little as an order of magnitude are thus effectively isolated from one another. Simon (1973) refers to this as "loose vertical coupling."

In addition to isolating levels from each other, differences in rates help explain how the system responds to environmental fluctuations. The fluctuations may be strictly periodic (i.e., describable by a sine or cosine curve) such as daily changes in light and seasonal changes in temperature, or they may be periodically recurring pulses. In either case, the characteristic rates of a level determine the frequencies of environmental fluctuations that can be attenuated, that is, the frequencies over which responses can, to some extent, be controlled.

Let us examine what all this means in terms of a biological system, the tree. Light striking leaves (lower level of hierarchy, faster time constants) fluctuates from day to day and even from minute to minute due to cloud cover and shading from other leaves. The leaves respond to these fluctuations by increasing or decreasing photosynthesis. If we examine the response of the higher level of organization, the tree, we find that oscillations are attenuated or dampened out. Thus, tree-ring width, reflecting the annual growth response of the tree, shows the integrated response to light changes. The high-frequency response of the leaves has been filtered out. Instead, the average or integrated response is seen in the growth increment at the level of the tree. The lower level of organization has communicated only its average response to the higher level.

Notice that this close connection between environmental

fluctuations and the characteristic rates of a level can only be made for linear systems. In general, nonlinear systems have a greater range of potential responses. However, whenever higher levels of the ecological hierarchy respond more slowly and therefore attenuate higher frequency fluctuations, the generalizations of the theory hold.

Horizontal Structure in Hierarchical Systems

Within an organizational level, a hierarchical system can be further decomposed into subsystems or holons (Koestler 1967, 1969) on the basis of differences in rates. Within each holon, components interact frequently (i.e., strongly) with each other but only infrequently (i.e., weakly) with components of other holons. Each holon can be defined in terms of a boundary or surface that encloses its components and separates them from components in the rest of the system (T.F.H. Allen et al. 1984). The holon is itself a whole composed of parts and is simultaneously a part of some greater whole (i.e., the next higher organizational level). We will reserve the word "component" to refer to the parts of the holon.

The surface of a holon may be visible and tangible, as the skin of an organism and the boundary of a lake, or intangible, as in the case of populations and species. However, even intangible surfaces can be described by the sharp gradient in rates that occurs as the boundary is crossed (T.F.H. Allen et al. 1984). Rates inside the surface characterize interactions among components and are relatively rapid and uniform. Rates outside the surface characterize interactions among complete holons and are relatively slow and weak. Thus, in biochemical studies, there is little problem in distinguishing the Krebs cycle, even though its components (i.e., chemicals) mingle with other contents of the cytoplasm. The boundary can be located by differences in rates.

Whenever a hierarchical level can be decomposed into

holons based on gradients in rates, the level is said to show "loose horizontal coupling" (Simon 1973). System behavior can now be explained in terms of interactions among these holons. By organizing characteristic rates into levels and holons, we have succeeded in abstracting a small-number system from the medium-number system.

Loose horizontal coupling is related to loose vertical coupling in the following sense. Rapid dynamics are isolated within the holon by its surface. Interactions among holons involve relatively longer time constants. The relatively slow interactions among holons become the dynamics of the next higher level in the system. The dominant time constants of each successive level are slower, and the holons at one level become the interacting components of higher-level holons. Thus, differential interaction rates result in rapid dynamics being isolated to one level (J. E. Schindler et al. 1980).

Within-level structuring also helps explain the relationship between macroscopic properties at one level and properties measured at some lower level in the hierarchy. Properties of components are averaged, filtered, and smoothed as they become part of the aggregated output of a holon. It is the aggregate output that forms the input to other holons and thence to higher levels in the hierarchy. The high-frequency detail in the aggregate signal is lost. A further filtering will occur at each level of the hierarchy. The greater the number of intervening levels separating two scales of interest, the less will be the recognizable influence of lower-level behavior on a higher level in the system. For example, T.F.H. Allen et al. (1977) discuss how difficult it is to explain successional changes in phytoplankton communities based on cellular physiology. The intervening levels of species, guilds, and strategies (T.F.H. Allen and Koonce 1973) attenuate the minute-to-minute changes in biochemistry and make it impossible to predict successional patterns based solely on physiology. Any attempt to relate a macroscopic

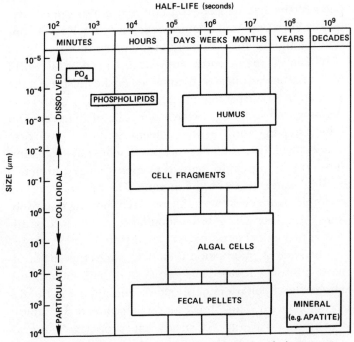

Fig. 5.2. The relationship between time and space scales in a plankton ecosystem (Scavia 1980).

property to the detailed behaviors of components several layers lower in the hierarchy is bound to fail due to the successive filtering.

Investigating the Hierarchical System

Now that we have seen how differences in rates can structure a system, we are in a better position to examine the problem of spatiotemporal scale introduced in Part I of the book. Ecological dynamics occur over a broad spectrum of space–time scales. Scavia (1980) points to phytoplankton–zooplankton interactions occurring at many scales from a meter to tens of kilometers (Fig 5.2). At each scale, some as-

81

pects of the system seem constant and others appear dynamic. Such apparent dilemmas are easily resolved by considering the system to be hierarchically structured.

Whenever the scientist approaches the natural world with a particular purpose in mind, he or she must select the scale appropriate for that purpose. Levin (1975) argues that defining and isolating the relevant scale of dynamics is a critical step in setting up any problem. Stommel (1963) provides a lucid discussion of processes that operate from minutes (gravity waves) to many centuries (ice ages) in determining water level of the oceans. He provides several examples from oceanographic research showing that problems result if one forgets the simultaneous influence of different processes operating at many different scales.

The scale of observation then determines that major attention must be focused at a particular organizational level. Higher-level behaviors occur slowly and appear in the description as constants. Lower-level behaviors occur rapidly and appear as averages or steady-state properties in the description. Thus, analysis of annual tree growth need not consider instantaneous changes in stomatal openings, nor long-term changes in regional climate. Simon (1973) argues that a system description is effective only if it permits a sealing-off of higher and lower levels of behavior.

From a hierarchical perspective, the definition of the system depends on the window (i.e., the range of rates) through which one is viewing the natural world. If one is looking at the effects of nutrients in a five-minute pulse of rain, the relevant components might be leaves, litter surface, fungi, and fine roots. Slower components will appear as background constants. On the other hand, if one is studying climatic changes over centuries, the relevant components might be large pools of organic matter. For changes that take place over centuries, hourly, daily, and seasonal

82

changes in fungal dynamics are smoothed out and can be ignored.

Gingerich (1983) shows that rates of morphological evolution measured over short intervals in the laboratory will be rapid because every minor change is observed and scored. Even if the evolutionary process is proceeding at exactly the same rate, measurements derived from long time intervals in the fossil record will indicate very slow change because a large number of changes will not be preserved in the record. It begins to become clear why it was necessary to emphasize space–time scales in Chapter 2. The interpretation of measurements taken on the system, and indeed the definition of the system itself, depends on the scale at which one is approaching the natural world.

Depending on the spatiotemporal scale or window through which one is viewing the world, a forest stand may appear (1) as a dynamic entity in its own right, (2) as a constant (i.e., nondynamic) background within which an organism operates, or (3) as inconsequential noise in major geomorphological processes. Thus, it becomes impossible to designate *the* components of *the* ecosystem. The designations will change as the spatiotemporal scale changes.

Sollins et al. (1983) analyzed soil organic matter accumulation at mudflows on Mt. Shasta, California. Their analysis exemplifies how the observable system dynamics are altered as one changes scale (see Fig. 5.3). Over centuries, organic matter is accumulating and major oscillations in the pattern of deposition result from fire-initiated secondary succession. On a finer scale of years, net accumulation is more difficult to detect and dynamics are due to annual litterfall and decomposition. If we take the analysis of Sollins et al. to an even finer scale (Fig. 5.3), annual dynamics dissappear as one focuses on wind-blown additions and removals and the action of large decomposer organisms. Thus, if one were attempting to explain soil organic-matter accumulation one

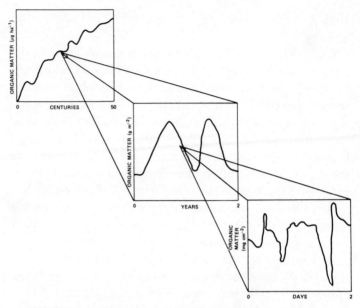

Fig. 5.3. Changes in apparent dynamics of a litter–soil system with a change in time and space scales. Slow dynamics over centuries show accumulation of organic matter with oscillations due to succession (Sollins et al. 1983). On a scale of years, seasonal decomposition processes are apparent, while an observation window of days reveals rapid fluctuations in litter due to wind and arthropods.

would come to a totally different description of the dynamics, depending on the scale being considered.

The hierarchical perspective makes it clear that the ecosystem is not simply the context for population and community dynamics. Such a viewpoint seeks to limit observations to a single window, a single observation set. Instead, behavior must be defined on the basis of the phenomena under examination. Hierarchy theory provides a consistent methodology for dealing with the natural world at many spatiotemporal scales.

The hierarchy viewpoint is helpful in deciphering the complexities of spatiotemporal scale outlined in Part I. For example, consider the effects of perturbation such as fire, windstorm, or precipitation. Ecosystem response must be explicitly tied to the temporal properties of the perturbation. There is not a monolithic ecosystem that responds in a single manner to all perturbations. The ecosystem is a hierarchical system that responds differently over a range of frequencies. If perturbations occur at a large scale, they selectively eliminate lower-level components unable to withstand the disturbance. However, higher levels see only the averaged or integrated responses of the components. As long as the perturbation does not occur at a scale larger than the entire system, there may not be a total disruption of function at the higher level of organization. The response is isolated at, and below, the level that responds immediately to the frequency of the perturbation. Higher levels are relatively buffered. The environmental perturbation can be defined in terms of its intensity, duration, and frequency of recurrence. These characteristics determine the level of organization that responds to this perturbation. Thus, momentary changes in CO_2 concentrations must be addressed at the level of stomatal openings and diffusion gradients in leaves. Monotonic changes in global CO_2 concentrations over decades and centuries must be examined in the context of the slowly changing major pools of carbon in the biosphere.

Thus, the ecosystem cannot be arbitrarily defined in space and time. Rather, the system must be defined relative to the scale of the problem being addressed. Short-term cycles of temperature and precipitation affect microbial processes. Seasonal and annual cycles influence the life-history evolution of organisms (e.g., Southwood et al. 1974). Iglich (1975) recognized twenty-five to forty-year cycles of reproduction in southern Appalachian forest associated with precipitation patterns of similar frequency. Lugo et al. (1976) suggested that mangrove forests are in phase with the frequency of tropical hurricanes. Successional dynamics of northern hardwood forests respond to a range of disturbances from short term (e.g., tree fall) to long term (e.g., major windstorms over broad areas) (Forcier 1975; Bormann and Likens 1979a, b).

Fire is an especially well-studied perturbation that becomes involved in the scale of system definition in a variety of ways (Ahlgren and Ahlgren 1960; Daubenmire 1968). R. F. Wright (1974) argued that fire frequencies provided the long-term stability needed to preserve conifer forest ecosystems. Fire is the perturbation that ensures landscape diversity and preserves seed sources for recovery from any major disturbance. Viewing ecosystems on the arbitrary scale of the forest stand results in seeing fire as a cataclysmic disturbance. This view led to a national campaign to eliminate forest fires. But viewed on a spatial scale appropriate to the frequency of recurrence, fire can be seen as necessary to retain the spatial diversity of the landscape and permit recovery from disturbance. Thus, the forest stand is not the proper scale on which to study large fires.

Summary

The picture that emerges is of a medium-number system whose organized complexity is amenable to decomposition because of differences in process rates (Simon 1973). Behaviors are arranged into a vertical structure because very slow and very rapid levels are isolated from each other. Within a level, holons are isolated from each other by gradients in process rates. Thus, the old imagery of the natural world as having everything connected to everything else is shortsighted. It is the relative disconnection that constitutes the organization of the system.

It is clear that ecosystems are level structures with dynamics over a broad range of interrelated space–time scales. The key to studying complex medium-number systems, such as ecosystems, is to make explicit their rate-dependent organization. As soon as we set sampling frequencies, we are placing observational windows on the system. If we choose an inappropriate scale of observation, the response of the system may totally elude us. The response may occur at a lower level of organization and appear only as a minor change in an average or integrated property (Overton 1977). On the other hand, the response may involve transport processes that cause changes in a portion of the landscape outside our chosen spatial scale and long after our temporal scale of measurement.

OTHER METHODS FOR DECOMPOSING A SYSTEM

At first sight, the emphasis on rates as the fundamental characteristic of the ecological hierarchy seems arbitrary. Certainly there are a number of ways that a collection of objects can be arranged into a hierarchy and there are a number of other criteria that one might use to decompose a system.

A question therefore arises as to whether any hierarchy, no matter how it is constructed, will fit the theory outlined in this chapter. The answer is no. A random assortment of objects can be assigned to groups on an arbitrary criterion (e.g., color) and these groups in turn assigned to larger and larger aggregates. The resulting system might be considered hierarchical, but our theory would be of little value in deciphering the system's behavior.

In general, it is insufficient to assume a hierarchical structure simply because components are aggregated. As an example, one could assemble nonlinear oscillators into pairs or larger groupings. The resulting higher-level aggregates do not necessarily show slower dynamics and may, in fact, respond with faster time constants than the isolated components (Gallup and Benson 1979). Thus, simply indicating that components are connected is not sufficient reason for assuming that all of the principles outlined in this chapter will fit.

Hierarchical Systems Based on Tangible Components

The rate criterion may seem arbitrary since so many ecological problems are addressed in terms of hierarchies formed of tangible objects such as individual organisms. Is not the criterion "tangible component" just as fundamental as differences in rate? The answer is that tangible boundaries are only a special case of boundaries defined on a strict rate criterion. Consider, for example, a thermister moved about in the body of a homeothermic mammal. We can use the temperature as an indicator of the rates at which metabolic processes are occurring. As we move the thermister from the core out to the skin, the rate processes change gradually. At the surface of the skin, there is a rapid change in temperature (i.e., a steep gradient in the rate processes). We can use this gradient to define the surface of the organism.

In the case of the mammal and its temperature, the boundary based on rate processes corresponds to the tangible, structural boundary that we detect with our human visual equipment. This is consoling because we use such structural boundaries in many ecological problems, focusing on the tangible organisms and its aggregates. However, we pointed out in Part I that because this method of defining components works well for one class of problems is no indication that this is the way one *must* define components. We saw that in many process–functional studies, insistence on tangible components led to problems.

If we maintain that components in an ecological system must be tangible in their own right, we have difficulty solving some functional problems of interest in ecosystem analysis. But, if we use differences in rate as our criterion, then we can include both tangible and intangible components. In fact, structural boundaries represent one class of discontinuities and arise where especially steep gradients in several distinct rate processes converge (Gerard 1969; J. R. Platt 1969). Thus, defining components based on discontinuities in rate processes can be seen as the more general approach (Gutman 1969). A number of authors (e.g., Overton 1977; MacMahon et al. 1978, 1981; Muller-Herold 1983) make the same point: levels in the biological or ecological hierarchy can most easily be defined by response time. As we will discover in the next chapter, modern theories of nonequilibrium thermodynamics (e.g., Nicholis and Prigogine 1977) lead one to believe that tangible boundaries emerge naturally from differences in rate processes.

In practice, discontinuities in rate processes are often used to define components; for example, consider the rhizosphere (Wilde 1968). The rhizosphere is the area immediately adjacent to the roots of terrestrial plants. It operates as a functional unit in the retention and uptake of nutrients. The rhizosphere is readily defined and measured using the rate criterion because rates of nutrients processing within it are rapid and relatively uniform. Its surface can be defined by the sharp gradient in exchange rates as one moves further away from the root. Therefore, the rhizosphere can be defined in terms of differences in rate.

On the other hand, defining the rhizosphere in terms of tangible objects is very difficult. The functional unit involves (1) a part of a tangible entity (i.e., root of the plant), (2) an abiotic component (i.e., soil water in the vicinity of the root), (3) a guild (Root 1967) of microorganisms, (4) complete tangible entities (i.e., mycorrhizal fungi living in intimate association with the root), and (5) members of other soil invertebrate populations, such as nematodes, that happen to be in the vicinity and that may enter and leave the functional rhizosphere at different times. Thus, the rhizosphere does not correspond in any simple way to an aggregation of tangible objects.

It appears that difference in rates is a more general criterion than tangible boundaries. In fact, this is a moot point because whenever tangible boundaries are appropriate, their convenience will recommend them. However, the insistence on the rate criterion is not simply semantics. The hierarchical levels properly defined on tangible boundaries, considered as a special case of differences in rates, will display the other properties we outline for hierarchical systems in the remainder of this chapter.

Hierarchical Systems Based on Spatial Criteria

An ecological system may also be decomposed on the basis of spatial discontinuities. The resultant organization is related to the frequency hierarchy, but the structures are not necessarily identical and a problem defined on space will not necessarily reduce to one defined on rate differences (e.g., Fig. 5.2 shows a spatial hierarchy at overlapping temporal scales). Nevertheless, the hierarchies defined on space and time share many properties. For example, the spatial hierarchy, like many rate hierarchies, is always nested because the higher (i.e., larger) levels contain and are composed of the lower levels. Ordinarily, the larger system responds more slowly, so the hierarchy defined on this criterion will often be related to the frequency hierarchy.

In some cases, size and frequency response may be very closely related. For example, Peterson et al. (1984) reviewed literature for forty-one species of birds and mammals. They found a significant allometric relationship between body size and the period characterizing population cycles in these species. As expected, the larger organisms showed slower frequency behavior.

Perturbations of a given frequency are often associated with a particular spatial scale in the natural landscape. Small forest fires occur frequently but over small areas. Fires that occur over larger areas have much longer recurrence times. In general, the lower the recurrence frequency of a perturbation, the larger the spatial scale that must be considered and the higher the organizational level of the system on which we must focus.

The watershed is an appropriate landscape unit for examining responses on the order of years to decades. However, for events occurring at intervals of 10^4 to 10^5 years, the system of choice becomes the entire deciduous forest. The biosphere becomes the focal system examining events on time scales of 10^6 to 10^7 years. Ecosystem boundaries are a matter of scale: spatial boundaries must be correlated with the temporal framework appropriate for a particular perturbation (MacMahon et al. 1978).

Hierarchies Based on Traditional Levels of Organization

Although we criticized the level-of-organization concept in Chapter 4, it is still true that hierarchies based on this concept are the most familiar to ecologists. Therefore, we must decide whether this type of organization fits the theory we are developing.

Some traditional hierarchies fit the rate concept nicely. Thus, organism/population/community seems a simple nested hierarchy in which each higher level is both slower and larger than the lower-level holons (MacMahon et al. 1978). The components can be defined both on the criterion of tangibility and on the rate process criterion.

However, in general, the hierarchies formed by the level-of-organization concept do not reduce to rate structuring. Take, for example, the assumption that ecosystems are a higher level of organization than communities. In fact, some community processes (e.g., succession or migration following glaciation) are considerably slower than some ecosystem processes. In addition, the spatial extent of a community, as defined by a species–area curve, may be larger than the spatial extent of the ecosystem appropriate for studying nutrient cycling.

In many respects, the level-of-organization concept is simply a way of arranging the many scales that are of interest to the ecologist. The smaller and "simpler" objects are placed lower in the hierarchy and the larger and more "complex" are placed at a higher level. The criterion for placing the ecosystem higher than the community is simply that the ecosystem analyst considers factors such as nutrient cycling in addition to the biota. Thus, the ecosystem is somehow more complex and should be placed higher. As a result, the hierarchy formed by the level-of-organization concept does not reduce to any simple or objective criterion for ordering. The arrangement is intuitive rather than operational and is based on an inadequate concept of complexity.

In the broadest sense, each of the traditional levels is itself a system that can be studied over many scales. Goodwin (1963) discusses scale for the cellular level of organization. The cell participates in processes that operate over seconds for chemical reactions and over geologic eras for evolutionary change. Simply stating that one wishes to address the dynamics of the cell is not a sufficient definition of the problem. One must also specify the scale of observation.

Thus, the traditional levels of organization are not strictly defined in terms of the spatiotemporal scales that are isolated at that level. It is not immediately obvious what spatial or temporal window is involved when one studies a population or community. Each of these levels operates over a broad spectrum of scales just as we have outlined for ecosystems.

The traditional concept remains useful for arraying the many complex systems emphasized in ecology. However, it is more a catalog of interesting problems than an opera-

92

tional basis for relating one scale to another. At the very least, it should be clear that when we discuss hierarchies in this and subsequent chapters, we are not referring to the traditional levels.

Summary

It appears that rate or frequency can be defended as a fundamental way of decomposing hierarchical systems. Tangible boundaries will be appropriate for many problems, but they are really only a special case of surfaces defined on differences in rate and all of the principles apply. Since space and time are interrelated, many spatial problems can be reduced to differences in rates of response. We must remain careful, however, because some spatial problems will not be reducible to the time domain and yet can be considered under the theory whenever the spatial organization is closely analogous to the temporal (i.e., whenever large systems respond to large perturbations in the same way that slow systems respond to low-frequency perturbations).

Not every definition of hierarchy will be suitable. Simply because components are partially connected does not mean that the system is hierarchical in the sense used here. Simply because components are aggregated in some way does not mean that the system will fit the concepts. Simply because one object can be designated as higher than another does not mean that our analysis is relevant.

RELATIONSHIPS AND ASYMMETRY

We have examined how differences in rate processes can be used to structure a complex system into levels and holons. We must now examine how this structure affects dynamic relationships among the parts. In particular we will be interested in the symmetry of the relationships.

A symmetric relationship is one in which X can affect Y and Y can affect X. The relationship is a two-way street. If all parts of a complex system interact directly and symmetrically, then there is a low probability that the system will endure (Gardner and Ashby 1970). A disturbance anywhere in the system tends to affect all other parts of the system. Therefore, a complex system must be organized so that asymmetric relationships structure the way in which the system responds to perturbations (T.F.H. Allen and Starr 1982).

Hierarchically structured systems will not be fully connected, and, due to differences in rate processes, some relationships will be asymmetric: X will affect Y, but Y cannot affect X. This asymmetry becomes a fundamental organizing principle. By taking advantage of asymmetry in rate processes, it is possible to dissect the dynamic organization of the system.

Symmetric Relations Are Interactions and Occur within a Level

On a given hierarchical level, the holons are operating at similar rates and can "interact," at least potentially. The relationships are symmetric. Thus, the individuals (i.e., holons) in a population (i.e., level) can affect each other directly through social interactions, competition, et cetera. The basic reason that this symmetry of relationships can exist is that each member of the interaction is operating on the same spatiotemporal scale. An action by one holon occurs on a scale that is geared to the spatiotemporal dynamics of the other holons on the same scale.

By focusing on symmetric interactions, it is possible to define structure based on the strength of interactions (Simon 1973). Some components within a level interact strongly, others only weakly. It is ordinarily possible to isolate components that interact strongly with each other. Thus, strongly interacting components form a specific holon and are delimited from other holons at that level. The strong interactions within a holon reflect fast rates of exchange between components. Fast rates of exchange may be seen as high-frequency activity within the holon. The rate change that occurs at a surface reflects the decoupling of components within one holon from components of other holons. The surface defines the separation of holons. In this way, the interaction between components is minimized (J. R. Platt 1969).

Asymmetric Relations Are Constraints and Occur between Levels

Asymmetric relationships occur between hierarchical levels and are called "constraints." Thus, higher levels can affect lower-level holons but are relatively unresponsive to changes in the lower level. The higher level appears as an immovable barrier to the behavior of lower levels. This constraint is a natural consequence of the asymmetry in rate constants. The rates always become slower as one ascends the hierarchy and, therefore, the lower levels are constrained because they are unable to affect the behavior of the higher level. Grene (1969) argued that hierarchical organizations actually involve a double asymmetry. Lower-level behaviors are essential to the functioning and persistence of higher-level structure that, in turn, constrains the behavioral flexibility of all lower-level objects (see also Mesarovic and Macko 1969).

The importance of constraint in ecological organization is easily seen in the contrast between laboratory and field behavior of organisms. In the laboratory, a holon (e.g., a population) is capable of a great range of behaviors. If we lift all possible limitations on the population (e.g., food, space, predators, etc.) the population increases at some maximum potential rate. Such a maximum growth rate would define what the population is capable of doing.

In contrast, the ecological system constrains what the population will actually do in the field. In the natural world, population growth rate cannot approach its maximum because of limited food, space, predators, and so forth. These constraints come from the ecological organization of which the population is a part. Specifically, the constraints come from the higher-level organization of the system.

Constraint is a fundamental property of hierarchical systems and helps explain persistence through time. If each population were permitted to grow as rapidly as it can in the laboratory, the system could become erratic. It is only when the behaviors of the individual holons are constrained that the system can operate within realistic limits. Such realistic limits are themselves constraints imposed by still higher levels of organization.

The concept of constraint leads us directly to the important idea of functional equivalence. Two components at a given level are functionally equivalent if the signals they transmit to the higher-level system are indistinguishable (Simon 1973; Conrad 1976). Equivalent components occur in the same subsystem, like species in a guild, but they differ in their tolerance to environmental conditions. Such a view is explicit in ecological theory, which recognizes the occurrence of species having similar functional roles and similar structural positions in networks of matter–energy processing (Root 1967; Cummins 1974; Hill and Wiegert 1980). As a result of this functional redundancy, the properties of the ecosystem are relatively insensitive to fluctuations in component species.

Asymmetry and Stable Manifold Theory

Much of the discussion on constraint relationships between hierarchical levels can be stated more precisely through the mathematical theory of manifolds. Basically the idea is that a complex dynamic system can be represented in an n-dimensional phase space. Each axis of this space represents a state variable of the system, such as a population in the community or a compartment in a nutrient cycle. The current state of the system is represented uniquely by a point in this space because this point represents the values of each of the state variables. The dynamics of the system are represented by the motion of this point in state space.

Fundamental to manifold theory is the fact that, under certain conditions, the dynamics of the n-dimensional system can be represented in a smaller, m-dimensional ($m < n$) space called the manifold. This is true if m of the system components are much slower than the others and if the system is stable to this manifold. In other words, if any of the components are moved away from this manifold, they asymptotically return.

An example of stable, constrained behavior is given in O'Neill (1971). O'Neill considered a model for a forest system with a radiotracer in the leaf litter. Soil arthropods, previously unexposed to the tracer, were then introduced. There was an initial transient as the radionuclide in the arthropods increased and came into balance with the litter (Fig. 5.4). The concentration in the litter slowly changed because of leaching and other losses. No matter what the initial concentrations in the arthropods, they move toward a trajectory that runs parallel to the slow changes in the litter.

The concept has been made precise in a theorem of Tikonov, translated and explained in a biological context by Plant and Kim (1975, 1976). Omitting some of the technical detail, the theorem divides the original n equations into m equations with slow time constants and $p = n - m$ equations with rapid dynamics:

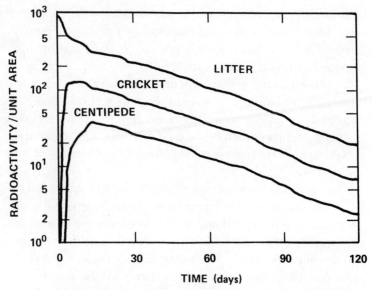

Fig. 5.4. An example of a stable manifold from an ecological model (O'Neill 1971). Slow changes in the litter component determine the overall dynamics of the system. Rapid dynamics in the cricket and centipede components appear initially, but subsequently the animals simply parallel the changes in litter.

$$du_i/dt = f_i(t, u_i, v_j), i = 1, 2, \ldots, m,$$
$$\epsilon_j \, dv_j/dt = F_j(t, u_i, v_j), j = 1, 2, \ldots, p.$$

Then, as ϵ_j approaches zero, corresponding to the fast components approaching a kind of equilibrium with the slower dynamics, the further dynamics of the system are dominated by the slower components. This is true as long as the system remains stable.

Basically, the theorem demonstrates that fast components of the system will rapidly come into balance with the slower components. Thereafter, the dynamics of the system will be dominated by the slow components, with the rapid components simply following along.

Translated into our hierarchical terminology, the slower

98

components belong to a higher level and impose constraints. As long as the system remains intact and stable, the next higher level determines the overall dynamics of the system. The lower hierarchical level can only operate within the constraints imposed. Thus, constraint in a hierarchical system is the necessary consequence of the difference in rate processes between levels. The hierarchical structuring of a dynamic system follows with mathematical necessity from the wide differences in rate processes. We will return in later chapters to this important theorem and its application to ecosystem problems.

CONCLUSIONS

The real advantage of hierarchy theory is that it offers an approach to medium-number systems that takes advantage of their organized complexity. In the past, we have decomposed ecosystems by assuming an a priori structure (e.g., populations or compartments) based primarily on human scales of observation. Hierarchy theory approaches the problem by searching for the rate structure that already exists in the system. Empirically derived differences in rates are used to decompose the system. The resulting structure is not arbitrarily imposed on the system; this organization is empirically derived from observations on the system.

This chapter develops a concept of the ecosystem as a hierarchical structure that responds to specific frequency windows of change in the environment. This approach is offered as an important tool for structuring our approach to the natural world. Chapters 1-4 showed that limiting study to a predetermined spatiotemporal scale (e.g., human perception and life span) or attempting to establish an arbitrary concept of constancy resulted in a distorted view of ecosystem dynamics and limited development of a general theory.

Ecological systems are enmeshed within and attuned to a dynamic environment that can be viewed as a rich spectrum of frequency signals. The system is organized into levels of organization that respond to, and continue in the face of, the full spectrum of changes. The level of the organization on which we must focus to explain and predict responses is determined by the intensity, duration, and recurrence frequency of the perturbation. Thus, the holistic concept of the ecosystem reduces to the identification of classes of behavioral frequencies, and leads to the development of sampling windows appropriate to measuring response. The emergence of such a view has been largely hindered by the tendency to view the natural world through human eyes (i.e., at a specific scale of resolution). It is becoming increasingly clear that we must view the world at the spatiotemporal scale at which it responds, rather than the space and time frame in which we operate.

100

CHAPTER 6

Hierarchical Structure as the Consequence of Evolution in Open, Dissipative Systems

In Chapter 3 we criticized approaches that were borrowed from engineering rather than being derived from observation of the natural world. We must now apply the same criterion to hierarchy theory to determine whether there is any basis for viewing living systems in general and ecosystems in particular in hierarchical terms.

The present chapter will develop the theme that all complex systems, including ecosystems, appear to be hierarchically structured as a natural consequence of evolutionary processes operating on thermodynamically open, dissipative systems (see also Wicken 1985). Because living systems are thermodynamically open and dissipative, our presentation should help establish the reasonableness of proposing hierarchy theory as a fundamental approach to ecosystems.

To accomplish our purpose, we will consider evolution in its broadest terms, as including physical and chemical evolution. Adopting this broad viewpoint will take us, at least temporarily, quite far from ecosystems. However, this approach is critical to establishing that thermodynamically open, dissipative systems at all levels of organization are subject to similar principles and seem to develop hierarchical structures as a result of these principles. Through a systematic, if brief, survey of evolution in physical, chemical, biochemical, genetic, and biotic systems we hope to establish that it is not at all surprising to detect the same hierarchical structure in ecosystems. To begin, we consider some principles of thermodynamics.

CONSERVATIVE AND DISSIPATIVE SYSTEMS

Modern physics came into its own in the seventeenth century when Newton formulated the laws of motion and gravitation. These laws completely describe the behavior of a mechanical system, given its initial conditions. Newton's laws were general enough to describe the orbits of planets, the trajectory of bullets, and the oscillations of pendulums. Of course, factors like wind resistance and friction created problems. But these were minor imperfections in a universe that otherwise behaved as a perfect clockwork.

Ignoring friction, a mechanical system is conservative with respect to energy. Once set in motion, the system stays in motion. In conservative systems, such as a frictionless pendulum, the energy can be transformed back and forth between kinetic and potential energy, but the total energy remains constant.

By the nineteenth century, experience with steam engines led to a generalization of the concept of energy to include both mechanical and thermal energy. The gradual slowing down of real systems could be attributed to the dissipation of mechanical energy into thermal motion of atoms. Total energy was conserved; it was merely transformed from one form to another.

This extension of the energy concept had only one less-than-perfect feature, although it was a very fundamental one. It was impossible to transfer all of the thermal energy into mechanical energy. In any system, the amount of thermal energy available for transformation into mechanical energy will spontaneously decrease through time. This is the message of the second law of thermodynamics, and entropy was introduced to quantify the idea.

Entropy always increases in an isolated system that has no energy or material transfers with an external environment. Consider two equal-sized containers of the same gas, one at a higher temperature, and both isolated from the external world. If the containers are connected by opening a valve, the molecules will mix and the entire volume of gas will reach a uniform temperature that is the average of the two initial temperatures. It can be shown that entropy has increased and reached a maximum when the gas has completely mixed.

The inevitable increase in entropy in the mixing gas has a probabilistic explanation. Once the two containers are connected, collisions between the fast molecules from the higher temperature and slow molecules from the lower temperature tend, on the average, to slow the former and speed up the latter. Also, there is spatial intermingling of the fast and slow molecules. Because entropy is an increasing function of the uniformity of a system, it inevitably increases or at least stays constant in a system with a statistically large number of molecules. Such probabilistic explanations lie behind all entropy phenomena. Hence, higher entropy is associated with more probable, and thus less structured, configurations of a system.

The second law of thermodynamics is often misinterpreted as meaning that entropy can never decrease and every system must eventually decay until entropy is a maximum and the degree of order a minimum. This is untrue because the law applies only to isolated systems. Many systems of interest are "open." An open system can exchange both energy and matter with an external environment. The total entropy change in an open system may be written

$$S = \Delta S_i + \Delta S_e \qquad (6.1)$$

where ΔS_i is purely internal and ΔS_e results from the interaction with the environment. Although ΔS_i must always be positive, ΔS_e can be positive or negative, depending on the nature of the interaction.

The Earth's atmosphere is a good example of an open system. Solar radiation provides energy that creates a global circulation of air. Without this source of energy, air currents would soon come to a complete stop (i.e., an equilibrium state of maximum entropy). The fact that the Earth's atmosphere remains in a state far from equilibrium is not a violation of the second law because it is an open system able to extract energy from solar radiation. The atmosphere can also be termed a "dissipative structure" because it maintains its structure (air currents) through the dissipation of solar energy.

Dissipative systems have an important property, called the minimum dissipation principle, that was first formulated by Onsager (1931) and later refined by Prigogine (1945, 1947). At steady state, a dissipative system produces minimum entropy per unit time compared to any nearby dynamic state (i.e., entropy production is a local minimum). Hence, this state can be maintained by a smaller input of energy than neighboring states.

The possibility of local decreases in entropy has been recognized as necessary to the existence of living organisms (Schrodinger 1945; Morowitz 1968; Bertalanffy 1975). Living forms are nonequilibrium, dissipative systems in which a high level of order is maintained by a flow of energy (Haskell 1940). In the case of autotrophs, energy from the sun is received as photons that raise chlorophyll and associated molecules (i.e., carotenoids, xanthophylls) to higher energy states. Heterotrophs are open systems with high-energy chemical compounds obtained from autotrophs and other heterotrophs. The metabolic activity of both autotrophs and heterotrophs dissipates this energy as thermal energy. This flux is sufficient to maintain organisms in a high state of internal organization, and thus to sustain life.

104

SELF-ORGANIZATION IN DISSIPATIVE STRUCTURES

Self-organization can arise in dissipative systems and can result in the hierarchical organization that characterizes living systems. While living systems are difficult to study in detail thermodynamically, many simple physicochemical systems have been analyzed. Even in fairly simple systems, external conditions can create spontaneous transitions from an initial state to a state in which energy dissipation takes place at a faster rate.

The French physicist H. Bénard experimented with a thin layer of oil that was heated from below while its upper surface was cooled by contact with the air. As long as the temperature gradient was small, nothing dramatic happened. The oil layer remained uniform, carrying heat from the bottom to the top by conduction. However, as the gradient increased above a critical value, the uniform surface suddenly changed to a tesselated pattern, usually a mosaic of hexagons. The hexagons indicated convective cells that now carried the heat. This new state was far from equilibrium.

The spontaneous jump to convective cells may be viewed as a process of self-organization. As the temperature gradient is increased, a point is reached at which the fluid is unstable. When a dissipative structure is near such an instability its entropy production reaches a relative maximum and it becomes sensitive to small fluctuations. A particular fluctuation can self-amplify until the uniform fluid changes into a highly ordered system of Bénard cells. Entropy production decreases again to a local minimum as this new stable state is approached. This phenomenon has been called "order through fluctuation."

Physical structures such as the convective Bénard cells are evanescent and will disappear when the external energy supply is removed. Other structures are more durable and can form building blocks for higher levels of organization. A plausible process by which complexity can build on itself has been called "stratified stability" by Bronowski (1973). As an example of this process, consider nuclear fusion in a young star that consists mostly of hydrogen nuclei. Occasionally two hydrogen nuclei collide and fuse into a helium nucleus, thereby releasing energy. This released energy is transformed into increased thermal energy of other hydrogen nuclei, which are then more likely to collide and fuse into more helium nuclei. This feedback eventually converts most of the hydrogen to helium. Similar fusion processes involve helium nuclei and then heavier nuclei. This "nuclear evolution" produced nearly all the elements in the periodic table with atomic weights up to about 60, the heavier elements being products of the runaway fusion reactions in supernova explosions (Hoyle 1977).

Chemical evolution subsequently used these atoms to build up complex molecular forms. Prigogine et al. (1972a, b) and Nicholis and Prigogine (1977) have speculated on the relevance of dissipative structures to prebiotic evolution. The energy provided by solar radiation into the original "soup" of inorganic compounds resulted in synthesis of simple carbon-based compounds. These simple compounds would have occasionally joined to form stable molecules of greater complexity.

As the number of complex molecules increased, competition would have occurred for the monomer building blocks. If a new autocatalytic process was more efficient at utilizing the monomers, it could replace some of the processes already present, causing the total energy dissipation to rise. In turn, each new state, more complex than its predecessors, could have made the system more vulnerable to new fluctuations. One might view this process as prebiotic natural selection, because it involves both the random ap-

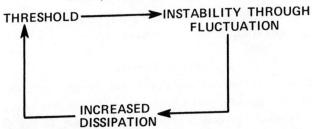

Fig. 6.1. The fluctuation–dissipation sequence as a feedback process. Macro-fluctuations produce instabilities that move the system to new organizational states. These new states permit increased dissipation and move the system toward new thresholds. New fluctuations can move the system beyond these thresholds, permitting the process to repeat itself.

pearance of new self-replicating "mutant" forms and competition between these forms for resources.

Prigogine et al. (1972a, b) pictured this fluctuation–dissipation sequence as a feedback process (Fig. 6.1). Each increase in complexity, triggered by a fluctuation to which the system is unstable, brings the system closer to a threshold above which the system can become unstable to new fluctuations. A long series of instability-induced transitions was necessary before this self-organization reached the level we call life.

THE LIVING SYSTEM EMERGES

Atoms formed in the interiors of young stars were the metastable building blocks on which early chemical evolution was based, and simple, metastable, covalently bonded molecules formed the basis for the evolution of more complex organic compounds. What made the higher hierarchical levels of life possible was the evolution of another level of stratified stability (Bronowski 1973): relatively stable molecules, or genes, that were capable of self-replication. These are not as stable as simple molecules, and certainly not as stable as atoms, but desiccation and freezing experi-

107

ments have shown that they are stable enough to preserve organization under adverse conditions.

Genes and Hypercycles

Bernal and others speculated that the first genes may have been crystals, perhaps silicates, rather than the nucleic acids that are the genetic basis for all life today. In any case, the early genes were probably direct-action genes (Cairns-Smith 1971), performing both replication and control functions. These genes probably resided in a primitive water or clay ecosystem with a solution of monomer building blocks from which the genes replicated themselves.

Mutations would have occurred from time to time, leading to a community of genes competing with each other for resources. As in modern ecological associations, mutualisms must have developed among some of these primitive genes. For example, gene type A might replicate better in the presence of type B because type B genes altered the environment. Groups of mutualistic genes, each specializing in particular functions, would have been efficient exploiters of their environment.

Eventually, the cooperative gene groupings became large enough to form their own internal environments. A threshold was crossed beyond which mutant genes that could not have existed on their own were able to survive in the sheltered environment. Coadaptations might have taken place until the individual genes could survive only in the environment of the group.

These primitive gene groupings (also called supergenes or phenotypes) may have resembled the catalytic hypercycles proposed by Eigen and his colleagues (Eigen and Schuster 1979; Schuster and Sigmund 1980; Eigen et al. 1981). These hypercycles are autocatalytic systems where each molecular entity plays a part in catalyzing the production of the next entity.

108

Eigen and his co-workers described the way hypercycles compete with each other. If two hypercycles share a molecular species, the outcome of the competition will depend on their relative efficiencies. If hypercycle 2 is more efficient than hypercycle 1, then the former will starve out the latter, inheriting the shared molecular species. This is a plausible model for prebiotic evolution.

Cairns-Smith (1971) subscribes to the orthodox view that takes primitive cell-sized globules as a point of departure. However, if the word "gene" is taken only to refer to one of Eigen's "species," then we can consider Cairns-Smith's argument in a more general context. He suggests that one of the early "genes" may have been ribonucleic acid (RNA). RNA probably performed some trivial function at first, but its structure facilitated the synthesis of proteins from amino acids. Protein is so useful as a building material that phenotypes with RNA would have a competitive edge. Only a few more steps would likely be necessary for deoxyribonucleic acid (DNA) synthesis by RNA. DNA became the master controller, replicating itself and serving as the template for messenger RNA.

Life is based on the establishment of polynucleotide–polypeptide cycles in a nutrient-rich medium surrounded by cell membranes. Polypeptides and other molecules produced by DNA can be used to form a cell wall, creating an internal environment sheltered from some of the vicissitudes of the external environment. These first cells, called procaryotic cells, maintained a controlled flux of energy and matter between their interior environments and the outside world and formed a new metastable building block.

Mutations as Fluctuations

In a community of procaryotic cells the same evolutionary process continues. P. M. Allen (1976) considered a procaryotic (asexual) population described by the equations

$$dx_i/dt = r_i x_i \left(1 - (x_i/K_i)\right) - \sum_{j=1}^{n} c_{ij} x_i x_j \quad i = 1, 2, \ldots, n. \ (6.2)$$

where

n is the number of genotypes, one for each of n alleles,

x_i is the number of cells of genotype i,

r_i and K_i are growth rates and carrying capacity of genotype i, and

c_{ij} is an interaction coefficient between i and j.

Assume this system is initially stable and that a mutant allele, $n + 1$, appears. What we must check is whether this new allele will be able to grow in the community. In the simple case where $n = 1$, the system of equations is:

$$dx_1/dt = r_1 x_1(1 - (x_1/K_1)) - c_{1,2} x_1 x_2$$
$$dx_2/dt = r_2 x_2(1 - (x_2/K_2)) - c_{2,1} x_1 x_2 \quad (6.3)$$

The situation can be represented by plotting a state plane (Fig. 6.2), assuming sufficient competition that the alleles can exclude each other. The states E_1 and E_2, where the entire population shares either allele 1 or 2, are stable. State E_3, where the two alleles could coexist, is unstable. Assume that the system initially is at E_1. For dominance to shift to allele 2, a fluctuation must move the system beyond the dotted line in Figure 6.2. Alternatively, slow changes in the system can cause the equilibrium to move to this point. After that, allele 2 will gain dominance along a trajectory that moves toward E_2.

Readers familiar with mathematical population ecology will recognize the above analysis. What P. M. Allen (1976) pointed out that was new is that the idea follows naturally from "order through fluctuation." In this sense there is a natural continuity between the evolution of prebiotic systems and the Darwinian evolution of biological organisms. To enhance this continuity we will continue the analogy be-

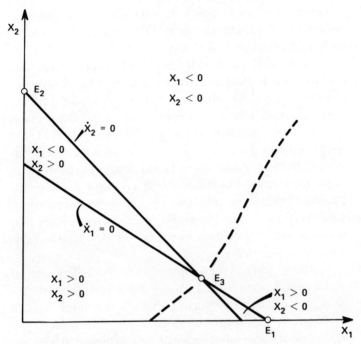

Fig. 6.2. Phase diagram for competition between two alleles, X_1 and X_2. At point E_1 only the first allele exists in the population. However, a fluctuation may move the system to the left of the dotted line. Subsequent dynamics will move the system to point E_2, with the second allele replacing the first.

tween fluctuations and mutations. In the following section, the word "mutant" will be used in a general sense for any form that is significantly different from the norm whether or not it is the result of a point locus genetic mutation (i.e., the technical definition of mutation in genetics).

Levels of Complexity

Evolution did not stop with the creation of procaryotic cells. Higher organisms consist solely of eucaryotic cells. In procaryotic cells, genetic material is not bounded by a membrane within the cell. In contrast, the eucaryotic cell centralizes most of the nucleic acid in the nucleus. In addition, eucaryotic cells contain other membrane-bound structures such as chloroplasts and mitochondria.

It is noteworthy that both mitochondria and chloroplasts contain some of their own unique DNA, capable of carrying out some, though not all, of their protein synthesis. This led to the suggestion that both mitochondria and chloroplasts derived from free-living organisms. Margulis (1971, 1981) suggested that the first eucaryotic cells may have been fusions of large anaerobic bacteria and smaller oxygen-respiring bacteria that subsequently evolved into mitochondria. Plantlike eucaryotes may have evolved from symbioses of blue-green algae and heterotrophic eucaryotes. Such mutualisms could create their own food and deal with increasing atmospheric oxygen.

The next level of development was the emergence of multicellular organisms. In these organisms, the cells became specialized to perform particular functions. No cell could survive and reproduce on its own. The pattern was again one of cooperative or mutualistic groups that were superior to individual cells in exploiting some resource.

Interesting questions about complexity arise here. Initially, one is tempted to ask how complex organisms could have evolved from simple abiotic compounds (Pattee 1972). But the real question is how ordered, cohesive, simple behavior of organisms could have arisen from a chaotic mixture of abiotic compounds. The integration of lower levels into higher levels of organization must necessarily involve some simplification. Each new level of organization involves self-simplification precisely because the organization imposes a new system of constraints that both decreases the complexity of behavior and ensures a new metastability that permits even further developments.

As Bronowski (1973) proposed, biological complexity seems to be arranged into discrete levels, with transition stages either rarely found or completely absent. "The more the study of the behavior and evolution of living organisms is extended, the more it shows breaks at each level of integration. . . . Each level of organization represents a threshold where objects, methods, and conditions of observation suddenly change" (Jacob 1976).

Beyond the multicellular organism, the next level of complexity involves mutualistic relationships among species. The evolution of animal pollination, for example, permitted angiosperms to displace more primitive flora over much of the Earth (Regal 1977). However, we must be careful at this point, because different criteria lead to different concepts as to what constitutes the next stage. There is not a singular arrangement of levels that takes precedence (see Chapter 4 and MacMahon et al. 1978). Mutualism between organisms is higher than metabolic interdependence of organism parts; but this does not mean that mutualism must be, in an absolute sense, the next level above multicellularity, for mutualism surely occurs between unicellular forms. Thus, our choice of mutualisms among species as the next level of complexity is not dictated by any absolute property of the natural world, but instead by a desire to understand the organizational structures of collection of organisms involved in ecosystems. Given this objective, it seems more useful to choose mutualisms among species as the next higher level of organization. Of central importance is the observation that adaptation is not simply a response to changing physical environments but may require coevolution involving many species.

Vermeij (1978) pointed out that the coevolutionary process promotes positive feedback and may result in instabilities. A species must adapt to the characteristics of other species in its community. If the species fits tightly into an established community, it becomes so dependent that it cannot exist outside of the community. A community com-

113

posed of species of this type is itself vulnerable to environmental changes. A fluctuation in climate, for example, could cause the loss of a few species and precipitate irreversible destruction of the entire interdependent community.

Are there communities that are so tightly interconnected that the loss of a few members seriously endangers all of them? Tropical rain forests and coral reef communities are probably the most internally cooperative ecological systems and have complex mechanisms for recycling nutrients. It is known that these ecosystems take much longer than other ecosystems to recover from a perturbation (Pomeroy 1970). Obviously, the delicate coadaptations present in these systems take considerable time to be restored following damage.

A number of authors have argued that coevolution, leading to higher levels of organization, need not involve intense one-on-one mutualism. Lotka (1925) pointed out that species did not evolve in isolation, but were a part of an evolutionary process which operated on the entire ecological system. Clarke (1975, 1976) asserted that herbivore diversity was not the result of a single species of predator, but of a collection of enemies that might change over time. Jaenike (1978) and Hamilton (1980, 1983) made similar arguments for the evolution of sex. Feeney (1975, 1982) and Jansen (1980) point out that plant chemical defenses evolved as a response to "herbivory" rather than to a specific population of herbivores. Levin (1983) considers that the selective feedbacks operating on a population are often "diffuse," i.e., mediated by processes at different levels of hierarchical organization.

IS THERE A DIRECTION IN EVOLUTION?

The theory of evolution relies on the concept of fitness as an explanation of evolutionary changes. One of the problems that resists simple explanation in this theory is the consistent trend toward increasing complexity (see Wicken 1985). We must address this problem if we are to consider food webs and communities as higher organizational levels that evolved naturally through the principles outlined in this chapter. To understand the problem, we must first consider what is meant by increasing complexity.

The Problem of Complexity

We have already introduced Pattee's (1972) counterintuitive insight that behavior tends to become simpler (i.e., more orderly) as structural complexity increases. Therefore, we must be careful in our use of the word "complexity." In particular, we must understand that complexity depends on how we choose to view the natural world. If we focus on the organism as a unit of observation, then the community appears to be very complex; that is, it is composed of a large number of our units of observation (organisms) and these units are interacting in a large number of different ways. But we can examine the community on other scales as well. If we were interested in world vegetation maps, we would focus on the community as our unit of observation. In this case, the community might appear as relatively simple, enduring in an orderly state over long periods of time and responding in simple and predictable ways to slow changes in climate.

Thus, complexity involves subjective criteria that depend on the scale of observation, and there is a sense in which increasing evolutionary complexity follows naturally from these criteria. It is not surprising that complexity increases, because details at the level of natural selection combine in a large number of ways to produce behavior at successively higher levels. T.F.H. Allen and Starr (1982) pointed out that complexity always appears when several levels of organization are called into play to describe behavior. Thus, it is important to realize that increasing complexity can be a

115

helpful organizing principle but one that is strongly influenced by the way we choose to observe the system.

An example drawn from ordinary experience will help make this point clearer. A printed electronic circuit is very complex. Its assembly is far beyond the capabilities of the hobbyist. On the other hand, assembling a radio from printed circuits is relatively simple although it requires some skill and patience. The hobbyist is far more likely to understand the radio in terms of amplifiers and filters than he is to understand the printed circuit in terms of quantum mechanics. Now the question is whether to regard the radio, composed of many printed circuits, as more complex than the individual printed circuit.

The relevance of the example is that one's judgment of complexity is dependent on the model one uses to link behavior across levels. Thus, a model that links thought processes directly to individual neurons will view thought as incredibly complex. The complexity arises because the model sees fine-grain entities, neurons, as the explanatory principle. Other models using an intermediate level of explanation, such as brain-regions or integrated neural networks, would not see thought as quite so complex.

This hierarchical perspective on complexity simplifies the problem of increasing complexity in evolution. As the investigator deals with successively higher levels (e.g., populations, guilds, food webs, communities), the distance between the level of observation and the level of explanation increases. The increasing difference in scale causes the observer to see considerable increases in complexity at each step. Looking at the problem in this way, tendencies for higher levels of organization to appear more complex is a direct result of observation.

Evolution and Complexity

After accepting these caveats on the use of the word complexity, we must still deal with the increasing organizational structure that has appeared since the original procaryotes. Some argue that the increase reflects a general law of nature. Skeptics object that complexity has only incidentally increased when the underlying mechanisms related to fitness have favored it.

In recent years, the theory of nonequilibrium thermodynamics has entered into the debates concerning evolutionary trends. Matsuno (1978) recalled Margalef's observation that ecosystems evolve toward a decreasing ratio of primary productivity to total biomass. Matsuno claimed that this is because the ecosystem is a dissipative structure in which the degradation per unit biomass would naturally be minimized.

Wicken (1980, 1985) and Saunders and Ho (1976) were primarily interested in the application of thermodynamic theory to increasing structure and complexity in the biosphere. Both attribute this increase to the increased efficiencies of complex systems in utilizing incident energy, although Wicken sees natural selection on the individual level and thermodynamics on the ecosystem level as complementary explanations. Saunders and Ho reject fitness as a mechanism for ecosystem evolution in favor of the thermodynamic explanation and as a result have drawn especially heavy criticism. Some critics argue that the validity of nonequilibrium thermodynamics outside of physicochemical systems has yet to be demonstrated. There is certainly, then, no compelling reason to believe that it is an appropriate way of looking at ecosystems.

Monod (1972) has argued forcefully for the dominant role of chance, rather than law, in the evolution of life. Monod sees structural complexity increasing through evolutionary time but as a result of chance mutation coupled with a selective process. A given mutation may make an organism either more or less complex. On the average, mutations resulting in increased complexity will confer greater fitness. Over long time scales, a trend toward increasing

A PROPOSAL FOR A THEORY

complexity will become apparent, though no law requires a change in that direction. Countertrends, such as structural simplifications in the case of parasites, are also theoretically possible, because decreases in complexity can confer greater fitness in specific situations. Monod's radical position has elicited responses from those who see a greater degree of directedness in evolution (e.g., Schoffeneils 1976). The difficulty of reconciling natural selection with increasing complexity can be further alleviated by observing that natural selection results as much from the biotic environment that species create for each other as from the physical environment. In stable environments, species may be selected more for compatibility with other species than for tolerance to the physical environment (Vermeij 1978). Primarily biotic adaptations can emerge that would be entirely unanticipated from the nature of the abiotic environment.

The important implication is that complexity can increase as new species find new ways of exploiting an environment that is becoming more diverse. Cornell and Orias (1964) hypothesized that the changing biota constantly create new potential niches. The process continues in positive feedback fashion, with the increasing stabilization of the environment allowing an increased diversity of species.

EVOLUTION OF HIGHER LEVELS IN
ECOLOGICAL SYSTEMS

The theory of thermodynamic dissipative structures, together with the idea of stratified stability (Bronowski 1973, 1977), provide an adequate basis for the development of still further levels of hierarchical organization. The theory of dissipative structures explains how the flow of energy can create new and highly ordered structures, while stratified stability describes how highly ordered structures can persist

118

and form building blocks for still higher levels of organization (e.g., Johnson 1981).

Thus, the mechanisms for producing new ecological levels are directly analogous to those that operated on the prebiotic level. Next, we wish to examine the extent to which the same principles can be applied to the ordering of communities. That is, do groups of species form stable subsystems that, in turn, compose larger subsystems, incorporating abiotic components along the way until the total ecosystem is reached?

Hierarchical Food Web Structure

To visualize how hierarchical structuring of food webs may have developed, imagine the immensely long period of Precambian time. Life at first consisted of primitive heterotrophs that fed on organic compounds created by physicochemical processes. A crisis occurred as the heterotrophs began to exhaust this very slowly replenishing resource. Fortunately, some heterotrophs evolved photosynthetic pigments and so avoided possible extinction by manufacturing their own food.

For eons, simple algae and bacteria inhabited the waters. Fossil records show vast tracts of ocean floor covered with stromatolites, or structures formed by single-celled blue-green algae. Probably only a few species of algae dominated wide areas (Stanley 1981).

The invention of predation changed this monotonous picture. Predators appeared about 800 million years ago, perhaps not long after sexuality evolved (Jantsch 1980). Enormous dynamic instabilities must have resulted when predators appeared on the scene. Whole regions may have been stripped bare of algae or bacteria, or may have experienced violent predator–prey cycles. Only with the evolution of autotrophs resistant to the predators could stability be regained.

The predators can be considered as life-threatening contingencies faced by the autotrophs. As new predators evolved, new strategies evolved to meet the threat, and hence greater specialization among the autotrophs. This reasoning led Paine and others to hypothesize that predation has played a primary role in the evolution of the great diversity of life. This hypothesis has promise, although Awramik (1971, 1981) has pointed out that stromatolite diversity actually declined sharply from about 680 to 570 million years B.P., probably due to the evolution and diversification of consumers.

Despite the continued instabilities in food webs as new species appeared, certain stable combinations of species were occasionally hit upon. Wilson (1969) termed these chance associations "assortive equilibria." Diamond's (1975) "permissible species combinations" are also a version of assortive equilibria (Simberloff 1981). In such combinations, modified or fine-tuned by subsequent coevolution, the balance of predation, competition, and mutualism was such that no member could squeeze others out. In addition, resources were used efficiently so that competitive invaders were kept out. These species combinations display, then, a higher level of stratified stability (Bronowski 1973, 1977).

What evidence is there that species communities can build themselves up by a method of stratified stability? Because field evidence bearing on this question is extremely difficult to obtain, large-scale computer simulations have been used as a convenient if simplistic surrogate. Recent studies modeled the assembly of ecological food webs as a process of predator colonizations and extinctions (Tregonning and Roberts 1979; Post and Pimm 1983). These studies employed systems of equations of the form,

$$dN_i/dt = f_i(N_i) \left[r_i + \sum_{j=1}^{n} a_{ij}N_j \right] (i = 1, 2, \ldots, m), \quad (6.4)$$

120

to which new species with randomly chosen parameter values were periodically added. The new species either failed to survive or successfully invaded the food web, pushing it to a new equilibrium. In these studies, the model communities gradually increased their ability to resist instabilities generated by new invaders. Thus, the groupings of species tended to form metastable groups as required by Bronowski's concept of stratified stability.

CONCLUSIONS

It appears that at every level of biological organization, the thermodynamic theory of dissipative structures plus the concept of stratified stability leads us to anticipate the evolution of hierarchical structures. Simon (1962) argued that all complex natural systems must evolve in this way. At each stage in the process, relatively stable subsystems must be formed that endure long enough to permit further stages of development. Both Levins (1973) and T. Platt and Denman (1975) elaborated similar views on the origin of biotic-level structures.

The existence of metastable subsystems has long been recognized in environmental systems. The guild, as a functional unit, is more constant, stable, and enduring than any of the individual species that comprise it (Root 1967). Only such relatively stable subsystems endure long enough for the next level of organization to evolve. Thus, the evolution of nutrient cycling requires the relative stability of the functional components of the recycling process.

A hierarchical view of ecosystems appears justifiable based on the analyses in this chapter. At each stage of development, the nature of the complex system leads, if not inevitably, at least statistically, toward new levels of organization. It does not appear that we are artificially imposing this hierarchical structure for the convenience of explanation. Instead, it seems that we must admit this type of organization as presented to us by our experience of the natural world. Thus, it seems reasonable to consider hierarchical organization as a fundamental principle that we can use to formulate an adequate concept of ecosystems.

Part IV

Applications of the Theory to Ecological Systems

In Chapters 3–5 we outlined the elements of hierarchy theory and proposed that it was useful to conceive of ecological systems as hierarchically structured. We also tried to show (Chapter 6) that it was not illogical to think that biological systems in general and ecosystems in particular acquired this hierarchical structure through evolutionary processes.

In the final section of the book (Chapters 7-9), we begin an exploration of the theory applied to ecological systems. In Chapter 7 we will consider how the theory helps elucidate the complex interactions among species populations in the community. In Chapter 8 we will apply the theory to problems dealing with mass and energy transfer in the ecosystem. The final chapter will then attempt to link the population–community and process–functional approaches. This synthesis is proposed as an integrative paradigm for the study of ecological systems.

These final three chapters try to show that the theory of hierarchies is more than heuristics. To demonstrate that hierarchy theory can serve as a useful scientific theory, we must do more than show its internal logical consistency or point to its intuitive appeal. We must look for both theoretical and empirical evidence that the biotic and functional aspects of ecosystems can be conceptualized as hierarchical.

123

We will also try to indicate that the theory can lead directly to testable hypotheses about ecosystem structure and function. In essence, this final section asks whether the general theory can be taken a step further and applied to the solution of ecological problems.

CHAPTER 7

Ecosystems as Hierarchies of Species

The population–community approach to ecosystems concentrates on species populations and their aggregates, such as food webs and communities. We have already seen in Chapter 6 that open, dissipative systems with stratified stability may be expected to evolve a hierarchical structure. The present chapter will explore to what extent the principles of hierarchy theory will yield new insight and understanding into the phenomena examined in the population–community paradigm.

PROBLEMS OF COMMUNITY COMPLEXITY

A series of studies beginning with Ashby (1954) showed that randomly assembled systems are less likely to be stable as the number of components increases. Gardner and Ashby (1970) showed a similar relationship as the connectance of the system increases. Connectance is the percentage of the possible component interactions that are non-zero.

Because of the importance of organization in a complex system, connectance may be a better measure of complexity than size. MacArthur (1971) argued that stable systems have an intermediate level of connectedness. The breakdown of organizational structure that precipitates unstable behavior can be caused by the system becoming either underconnected or overconnected.

The type of change due to underconnectedness is exemplified by experiments on keystone species (Paine 1966). In such experiments, a keystone predator is removed from the system. Prey populations begin to grow rapidly and exhaust their food supplies. This sets off a chain reaction in which populations are eliminated from the system and the constraint system is changed. Following the exponential growth of some populations and the disappearance of others, the system tends to settle into a new stable configuration, rather different from the original state. T.F.H. Allen and Starr (1982) provide a number of additional examples in which underconnectedness leads to instability.

At the other extreme, the system may become unstable due to overconnectedness. This is the kind of instability investigated by Gardner and Ashby (1970). For large systems, they showed that increasing the number of direct connections among components decreased the probability that the system would be stable.

Focusing specifically on randomly assembled food webs, May (1972, 1973a) showed that the probability of instability increases with food web size and the average strength of interactions. In fact, if food webs were randomly organized, nothing the size of a natural food web could possibly be stable. One resolution of this paradox, as May realized, was that real food webs are not randomly assembled, but are subject to organizational constraints. May reiterated Margalef's (1968) observation that "species with strong interactions are often part of a system with a small number of species." In this section, we will review theoretical studies indicating that hierarchical organization can account for stability in complex systems such as food webs and communities.

In terms of hierarchy theory, overconnectedness occurs when the lower-level entities in the system begin to interact too directly. Contagious infection or pests in an agricultural system are constrained by a low level of contact in a diffuse population. But in a dense population, disease can run rampant. MacArthur (1972) considered two prey species populations in competition and showed that larger competition and predation terms increase the likelihood that one or the other prey population will be driven to extinction. Levins (1974) argued that the more connections there are in a system, the greater the chance that an unconstrained positive feedback will emerge to destroy the system configuration. May (1973b) pointed out that the more complicated a system (i.e., the more components there are with direct interactions) the more there is to go wrong. If everyone is trying to constrain everybody else, the system collapses. The essential problem is that diversity per se has nothing to do with stability. It is indirectly associated with stability, but through a complex relationship with connectedness.

Because there are two ways that a system can become unstable, by being either overconnected or underconnected, the addition of a new component that raises diversity can have opposite effects depending on the circumstances. An increase in diversity can make the system stable either by adding connected components to an underconected system or by adding disconnected components to an overconnected system. Conversely, there are two ways that increased diversity can decrease stability: by adding a highly connected component to an overconnected system, or by adding a disconnected or weakly connected component to an underconnected system.

If we are to elucidate the relationship between complexity and stability, it will be necessary to return to earlier insights into system organization (e.g., Odum 1953; MacArthur 1955). In the case of complexity–stability, it may be the connectedness of the system rather than the number of components that is most important in determining stability. Hierarchy theory suggests that in a stable system, direct and symmetric connection should be isolated in well-defined holons. Interactions between components of different holons should be rare. The system should be further stabilized by constraints representing asymmetric relationships between levels in the hierarchy. Thus, the theory has implica-

127

tions for both the level of connectedness to be found in a stable ecological system as well as the symmetric structuring of the connections. The theory shows promise of permitting an approach to the complexity–stability problem that takes better cognizance of the process-rate organization of ecological systems.

If a multispecies community is hierarchical, it should be arranged into weakly interacting holons of strongly interacting species. Indeed, many studies (e.g., May 1972; McMurtrie 1975) show that a complex system need not be unstable if it is organized in this manner. Levins (1970), for example, experimented with complex models in which a "secondary simplicity arises" because components tend to group into subsystems with similar rate constants (Levins 1973).

Theoretical studies indicate that organizational structure determines whether or not a system will be stable when species are added or deleted. Thus, adding a random component to a system that is already stable is more likely to result in stability than assembling the same number of components *de novo* (Makridakis and Weintraub 1971a, b; Makridakis and Faucheux 1973). These same studies indicate that the stability of a system is enhanced if the new component tends to structure the system into subsystems.

A recent study of some relevance is that of Roberts and Tregonning (1980) who considered model species obeying the equation

$$dN_i/dt = f_iN_i(r_i + \sum_{j=1}^{n} a_{ij}N_j) \qquad (7.1)$$

where

N_i is the population size of the ith species,
r_i is its intrinsic rate of increase, and
a_{ij} is an interaction coefficient between species i and j.

In particular, Roberts and Tregonning were concerned with the feasibility of random assemblages of this form. Feasibility means that all values of the equilibrium, N_j^*, have positive values. Mathematically, the population sizes at equilibrium may just as easily be zero or negative but such equilibria hold little interest for the ecologist. What does feasibility imply about all possible subsystems that can be constructed by removing one or more species?

To answer this question, the authors first constructed feasible assemblages of twenty-five species with interaction coefficients:

$$a_{ij} = \pm 0.16$$
$$a_{ii} = -1$$
$$r_i = \pm 1,$$

where the assignment of + or − was random.

From this feasible system, q species were randomly eliminated, and the new systems of $m = 25 - q$ species were sampled and tested for feasibility. The same process was performed with an initially unfeasible system of twenty-five species. The results indicate that an initially feasible system is more likely to contain feasible subsystems. For example, with $m = 16$, 6 percent of the initially feasible systems remained feasible while none of the initially unfeasible systems was feasible. The authors interpreted their findings as evidence of the robustness of natural systems. They theorized that natural systems may consist of viable nested and overlapping subsystems, implying a "defense in depth" against collapse.

Recall from our discussion of the watchmakers, Tempus and Chronos, that Chronos constructed watches of 1,000 components from stable subassemblages of 10 components (see Chapter 4). Did Roberts and Tregonning discover a similar principle at work? Not exactly. They showed that a feasible system often consists of feasible subsystems which may be nested within each other or have overlapping species. This is as if Chronos had an enormous variety of ways of putting together his watches so that all intermediate stages were stable, but many of these ways might not be so straightforward as his independent subassemblages of 10 components. Thus, the specific organization found in the simulation, while still hierarchical in structure, was not as simple as might be found in mechanical devices.

The differences between hierarchies of parallel, nested, and overlapping subsystems are shown in Figure 7.1. Roberts and Tregonning (1980) emphasize the last two types, but certainly do not exclude the first. Considerations of simplicity recommend the first, or parallel, subsystem structure. The idea that a complex community structure may be built from simpler stable parallel subsystems is an attractive one. If the couplings between the subsystems are relatively weak, then the community may be approximately decomposable into independent units of a small number of species each.

CONNECTIVE STABILITY

It follows from the discussion in the preceding section that some type of hierarchical structuring is probably necessary for a community to be feasible and to be stable in the Lyapunov sense (i.e., the population sizes tend to return to equilibrium following a small perturbation). In this section we will focus on the types of hierarchical structure that favor community stability. One way of probing community structure is to ask what happens to a stable community when one or more species is removed. Is the residual system stable in the Lyapunov sense? Siljak (1975) terms this kind of stability "connective stability."

Analytical Studies

The importance of specific community organization is reflected in studies of Lyapunov stability and connectance

130

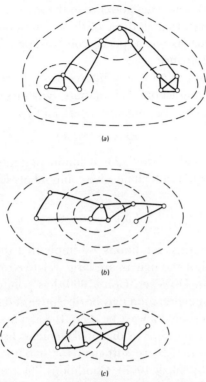

Fig. 7.1. Three potential hierarchical structures for communities. The species (circles) may interact (lines) as though they were organized into independent modules (a), nested into more and more complex associations (b), or developed into overlapping modules (c).

(DeAngelis 1975a). DeAngelis considered a food web with three trophic levels and ten species and asked what conditions would reverse previous results and favor an increase in stability with increasing connectance. He found that stability increased in three cases: (1) the consumers were inefficient, i.e., the ratio of biomass assimilated is a small fraction of the biomass removed from the prey; (2) the higher trophic-level species experience a strong self-dampening force that controls their population growth; and (3) there is a bias toward donor dependence in the interactions. It can easily be shown that all three of these conditions force the interaction matrix into a specific configuration called quasi-diagonal dominance.

Let us consider an $n \times n$ matrix $A = (a_{ij})$ representing community interactions. Community dynamics in the vicinity of an equilibrium point are described by

$$d\underline{X}/dt = A\underline{X} \qquad (7.2)$$

Matrix A is called quasi-diagonal dominant if there exists an n-vector, $d = (d_i)$ with all positive components such that

$$d_j|a_{jj}| > \sum_{i=1}^{n} d_i|a_{ij}|, \quad j = 1, 2, \ldots, n \qquad (7.3)$$

In simplest terms, conditions 7.3 imply that the terms on the diagonal of the matrix are large relative to the off-diagonal terms. However, Goh (1980) has pointed out that this simple interpretation may be deceptive and strict application of 7.3 should always be used. If A is a negative diagonal matrix (i.e., all of its diagonal elements are negative), then conditions 7.3 are sufficient for stability (McKenzie 1966, cited by Siljak 1975). Conditions 7.3 are not necessary, however. This is important because, as we will see, certain systems like the intertidal community of Paine (1966, 1969) probably do not obey these conditions.

Conditions 7.3 are equivalent to another important set of conditions. Suppose the off-diagonal elements of A are replaced with their absolute values to give the matrix A^*. Quasi-diagonal dominance is equivalent to the qth principal minor of A^* having the sign $(-1)^q$ for all values of q from 1 to n. These conditions are known as the Sevastyanov–Kotelyanskii inequalities and simply state that successive principal minors alternate signs.

The Sevastyanov-Kotelyanskii inequalities have an important implication for hierarchical community structure.

132

If the principal minors alternate signs in the n-species system then they will also alternate for the $(n - 1)$-species system, the $(n - 2)$-species system, and so on. Thus, the sufficient conditions for stability of the n-species are also sufficient when one or more species is removed. Siljak (1975) was able to demonstrate this rigorously by showing that any combination of off-diagonal elements could be set to zero and the matrix would remain stable. In other words, within each stable system are nested stable subsystems of all lower orders. Since these conditions are sufficient but not necessary, they do not imply that the nested subsystems *must* all be stable for the n-species system to be stable, but the implication is that a tendency toward this type of nesting helps guarantee stability of the total system.

One limitation of this result is that it does not take into account that species removal inevitably alters the position of the equilibrium point by changing the density of all remaining populations. It may, in fact, cause one or more populations to become mathematically negative, for example, if a key autotroph were removed, one or more heterotrophs might become extinct. Goh (1980) amended Siljak's (1975) theorem by requiring that the new equilibrium be feasible.

The results presented above hold only for negative diagonal matrices, that is, for cases in which each species has a net tendency toward self-regulation at equilibrium. This cannot be expected to be true in real communities. However, it is possible that a real community may consist of stable subsystems or modules, in each of which there may be internal self-regulation, although some subsystems may be regulated by exogenous factors, say, predators. Siljak (1975) showed that each such module may be represented by a single variable. A matrix made up of such modules will have all of its diagonal terms negative. Hence, connective stability can emerge as higher-level stability of many interconnected stable subsystems.

Simulation Studies

The analytical work of Siljak (1975) gives important insights into the hierarchical structure of communities. It cannot, however, tell us anything about the dynamics of communities following deletion. This can only be done by dealing with models in the global domain, where nonlinearities may be important. Computer simulation is necessary to study most of the global properties of nonlinear models.

Pimm (1979) undertook an extensive simulation study of species deletion stability. By species deletion stability, Pimm meant the tendency for a residual system to be both feasible and stable following species removal. Beginning with model systems having stable, feasible equilibria, Pimm deleted each species in turn. The resulting models were then checked for feasibility and local stability. Pimm used both Lotka–Volterra and linear, donor-dependent models in which only predator–prey relationships were present.

Pimm showed that species deletion stability depends on the model. Linear, donor-dependent models are stable following removal of top predators. This is simply because the flow of effects is one-way, from prey to predator, in donor-dependent models. When Lotka–Volterra equations are used, however, instability can result from removal of species from both the lower and upper ends of the food web. When plants and herbivores were removed, more complex food webs were more stable. When predators were removed, however, further species losses were more common if their prey were polyphagous.

QUALITATIVE STABILITY AND HIERARCHICAL STRUCTURE IN FOOD WEBS

In ecological systems the qualitative nature of interactions (e.g., predation or mutualism) are often understood even though numerical values cannot be assigned to the strength of the interactions. It is natural then to examine how system characteristics such as stability are related to purely qualitative aspects of system structure. In this section we discuss the concept of qualitative stability, which was introduced to ecology by Levins (1973) and May (1973b).

In an earlier section we discussed the Jacobean matrix A associated with a food web model. For a food web model to be stable at an equilibrium point, it is necessary and sufficient that the real parts of all eigenvalues of the Jacobean be negative. The eigenvalues are roots of the characteristic equation,

$$|A - \lambda I| = 0 \tag{7.4}$$

where the expression within bars denotes the determinant. Equation 7.4 can be expanded in a form that appeals more directly to biological intuition,

$$\lambda^n + \sum_{k=1}^{n} F_k \lambda^{n-k} = 0 \tag{7.5}$$

where $F_k = (-1)^k(-1)^{k-m}L(m,k)$, and $L(m,k)$ is the product of k links (each coefficient a_{ij} is one link) that form m disjunct feedback loops. For example, consider a predator feeding on two prey which compete with each other (Fig. 7.2). The $L(m,k)$ for this system are

$$L(3,1) = a_{1,1} + a_{2,2} + a_{3,3}$$
$$L(1,2) = a_{1,2}a_{2,1} + a_{1,3}a_{3,1} + a_{2,3}a_{3,2}$$
$$L(2,2) = a_{1,1}a_{2,2} + a_{1,1}a_{3,3} + a_{2,2}a_{3,3}$$
$$L(3,3) = a_{1,1}a_{2,2}a_{3,3}$$
$$L(2,3) = a_{1,1}a_{2,3}a_{3,2} + a_{2,2}a_{1,3}a_{3,1} + a_{3,3}a_{1,2}a_{2,1}$$
$$L(1,3) = a_{1,2}a_{2,3}a_{3,1} + a_{2,1}a_{1,3}a_{3,2}$$

Notice that the F_k in Equation 7.5 represent the sum of the feedbacks of length k. Thus, for feedbacks of length 1 we would have:

$$F_1 = (-1)(+1)L(3,1) = -a_{1,1} - a_{2,2} - a_{3,3}.$$

135

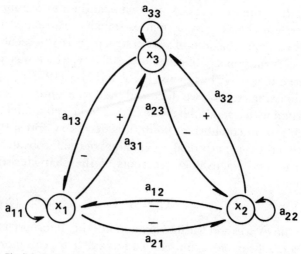

Fig. 7.2. Qualitative interaction structure for a predator, X_3, feeding on two competing prey species, X_1 and X_2. The small loops, a_{ii}, represent self-regulation and are assumed to be negative.

The reason for formulating the stability criteria in terms of feedback loops is that it is possible to show (May 1973b) that if there are certain qualitative limitations on the feedback loops, the Routh–Hurwitz criteria for Lyapunov stability of the system are satisfied. The limitations are:

(i) $a_{ii} \leqslant 0$ (for all i)

(ii) $a_{ii} \neq 0$ (for at least one i)

(iii) $a_{ji}a_{ij} \leqslant 0$ (for all $i \neq j$)

(iv) $a_{ij}a_{jk} \ldots a_{ii} = 0$ (for sequences over two)

(v) $\det(A) \neq 0$

What these criteria mean is that (i) no first-order feedback loop is positive and (ii) at least one of them is negative; (iii) no second-order feedback loop is positive, and (iv) there can be no feedback loops of lengths greater than two. Condition (v) simply forbids singularities that might be caused, for example, by the presence of two or more identical species.

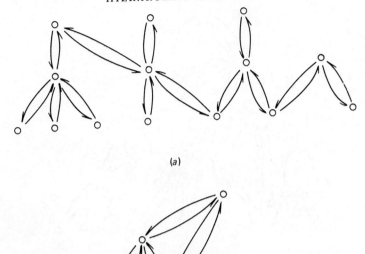

(a)

(b)

Fig. 7.3. Interaction structure for two communities. For purposes of stability analysis, it is important to distinguish between communities that contain no feedback loops greater than length two (a) from those that contain longer loops (b).

The feedback criteria place certain restrictions on the shape of the food web. The criteria are satisfied by systems like Figure 7.3a, but not for those like Figure 7.3b that have loops of length greater than two. This result relates to our earlier discussion that overconnection is one possible cause of system instability and that a lower level of connection, with hierarchical organization, can be stabilizing.

Tansky (1978) defines a branched-chain structure as one composed of those species that are mutually connected by k line segments in a branched-chain form without forming any loops. Figure 7.3a is such a structure. This contrasts with loop structures, in which we can return to the same

137

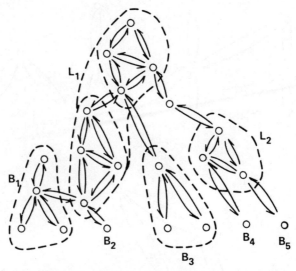

Fig. 7.4. Interaction structure in a complex food web. The species interactions can be grouped into loop structure groups (Tansky 1978) indicated by dotted lines and labeled L_i and B_i, Notice that all feedback loops of greater than two steps can be isolated within the loop structure groups for this system.

species along the connecting line segments, passing each segment only once. Figure 7.3b contains a simple loop structure involving three species.

It would provide great simplification if all ecological communities were structured in branched-chain form. One would then automatically be satisfying criterion (iv). Of course, we know that this is unrealistic and that nature is much more complicated. However, Tansky (1978) has taken an important step in generalizing the criteria to cases in which loop structure groups (LSGs) exist. All loops in the system are isolated within these subsystems. Then the total system may be stable if its overall structure is a branched chain with the LSGs considered as points (e.g., Fig. 7.4). In essence what Tansky has shown is that the stability of the overall system can be quite insensitive to the fine structure at lower levels in the hierarchy as long as that fine structure is contained within an LSG. This is precisely the same conclusion we reached in Chapter 5 relative to the sensitivity of overall system dynamics to fine structure at the next lower level as long as that fine structure was contained within a holon.

COMPUTER STUDIES

Computer simulations have been helpful in two important ways in the study of food webs. First, they have inspired new ideas and new research directions. For example, the intense work on stability–connectance in food web models was stimulated by Gardner and Ashby's (1970) computer studies of randomly generated matrices. Second, computer simulations have been a means of extrapolating our knowledge from simple systems to systems of complexity far beyond the limited applicability of analytic mathematical methods. We have already mentioned some results on the feasibility and stability of systems. Here we discuss some additional significant findings.

McMurtrie (1975) looked for relationships between size, connectance, and hierarchy in large, complex systems. His point was that particular patterns of organization might be more stable than purely random assemblages. Like May (1972) and others, McMurtrie considered the linearized Equation 7.2. McMurtrie chose the coefficients from Gaussian distributions as follows:

a_{ij} with mean zero and variance s^2
a_{ii} with mean -1 and variance s^2.

Various levels of connectance, c = the fraction of nonzero connections, were chosen, with the remaining $1 - c$ connections set to zero. We will consider three of McMurtrie's systems with $n = 24$ species (Fig. 7.5). The first

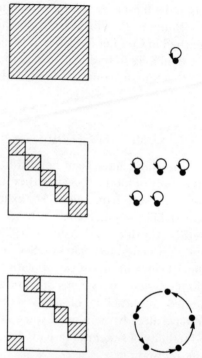

Fig. 7.5. Three classes of interaction matrices considered by McMurtrie (1975). The top case is a completely random organization and any species may interact with any other. The middle case is subdivided into a set of independent subsystems. The bottom case is also divided into subsystems, but the subsystems interact with adjacent subsystems.

(Fig. 7.5a) represents a completely random structure. In the second (Fig. 7.5b), the community is broken into several completely isolated subcommunities. In the third (Fig. 7.5c), a hierarchy of interactions is assumed with subcommunity i interacting only with $i-1$ and $i+1$.

McMurtrie performed twenty-five, fifty, and twenty-five simulations for each of these structures, using $c = 2/24$ and $s^2 = 0.5$ for each case. The results were as follows: (a) 16/25 stable, (b) 4/50, and (c) 23/25.

140

One of these results, b, is easy to interpret. With several independent subsystems, the instability of even one subsystem is enough to cause the system to be classified as unstable. Let us paraphrase Pimm's (1982) relevant discussion of compartmentalization. Imagine twelve species structured with random interactions such that the critical connectance for stability is 1/2. If the species are organized into three independent compartments of four species each, the number of interactions within compartments must be increased to keep the same connectance as the twelve species system. So each compartment must have connectance 132/36 times higher than the homogeneous system's 1/12; that is, 11/36. Now the critical connectance for a single subsystem is 1/4. So the subsystems are less likely to be stable, and the compartmental system is less likely to be stable than the homogeneous one. The reasoning probably applies to McMurtrie's case (b).

If the subsystems are linked as in Figure 7.5c, without adding appreciably to the number of cycles in the whole system, the mutual effects of the subgroups on each other may be stabilizing. This hierarchical structure (c) was shown by McMurtrie (1975) to be more stable than the random homogeneous structure.

One may question whether the type of structure shown in Figure 7.5c will be observed often in nature. We might more often expect a branched-chain structure (Fig. 7.3a). But, intuitively, it seems that this latter structure will also, on the average, be more stable than the random assemblage.

Another important contribution to the simulation of food webs is that of Austin and Cook (1974). They simulated food webs of eight, thirty-four, fifty-one, and sixty-one species, plus several organic pools, and recorded species extinctions. The main innovation of these authors was to include two important features of biological complexity that simpler community models leave out. (1) Predators can switch feeding preferences during the simulation to meet target feeding rates. (2) Feedback loops in the system occur because of predation on decomposer species.

The general result was that the systems were much more stable than would be expected from May's (1972, 1973a) stability–connectance results. The hierarchical structure of the systems, with the top predator's switching ability, contributes greatly to this stability. However, limit cycles are considered stable in these simulations. Austin and Cook (1974) also noted that there is an increase in the number of stable equilibrium points, or distinct stable limit cycles with increasing species numbers. In the more complex systems, few perturbed systems return to the same limit cycle.

EMPIRICAL EVIDENCE OF HIERARCHICAL STRUCTURE

What field evidence is there that food webs are compartmentalized into tightly interacting subsystems or compartments, each with only a few species, that interact only weakly with each other? A number of studies have addressed this question, either directly or indirectly, and we must examine the results to see if hierarchical concepts are applicable to ecological systems.

Recognition of the fact that many species play similar functional roles has led to attempts to define subsystems on this basis. Root (1967, 1975) has established that insect communities are often organized into guilds. The members of a guild have similar functional roles leading to strong competitive interactions within the group. Thus, although some members of the guild can be expected to appear in a community, the particular species are determined by their competitive advantages under the specific circumstances. Within the guild, competitive interactions can be strong. Between guilds, differences in functional roles tend to separate niche spaces and lead to weak competitive interactions.

Based on a similar line of logic, Cummins (1974; Cummins et al. 1973) subdivided benthic communities in streams into functional groups. Each functional group, such as grazers, shredders, or filter-feeders, is composed of species populations that can be expected to show strong competitive interactions. Once again, differences in function lead to weak competitive interactions between members of different functional groups. The designation of functional groups seems to capture something significant about the organization of these benthic communities and has influenced stream research and theory (Boling et al. 1975).

Organizational states above the species level have been found using multivariate analysis of phytoplankton data (Allen and Koonce 1973). The species tend to be aggregated into distinct subsystems based on their physiological tolerances. Thus, tolerance of low temperatures and need for high nutrient concentrations define an early spring grouping, while preference for higher temperature and tolerance of low nutrient concentrations define a late summer assemblage (Allen and Koonce 1973). Within these assemblages competitive interactions are strong and in some years individual species will not be detected in field samples. However, the groupings themselves are distributed as discrete units over the growing season. Although the groups overlap, the seasonal separation results in temporal isolation and only weak interactions between assemblages. Similar subsystems of phytoplankton have been found in simulation models of lake ecosystems (O'Neill and Giddings 1979) and in microcosm experiments (Waide et al. 1980).

Murdoch (1979) pointed out that some of the most carefully studied communities, for example, rocky shore intertidal systems, have a high proportion of strongly interacting

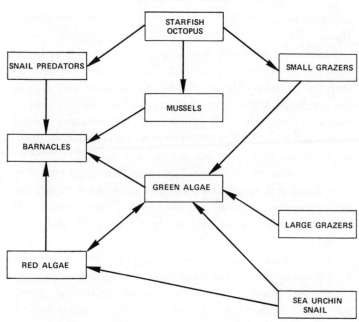

Fig. 7.6. Diagrammatic representation of the intertidal algal community studied by Sousa (1979). Experimental removal of any species in this system results in marked changes in the green algal component and in the community in general.

species (Connell 1970; Paine 1974; Sousa 1979). He discusses Sousa's experimental manipulations of a tidal algal community in particular (Fig. 7.6). Removal of any one species resulted in marked change in the dynamics of Ulva (a green alga) and in the community in general. Murdoch's interpretation was that, in a given ecosystem, it may be difficult to sort out subsystems that will be independent of strong interactions outside the subsystem.

It is interesting that Paine (1980) draws somewhat different conclusions. He asserts that strong interactions can encourage the development of subsystems. The individuals in these subsystems may be coadapted and exhibit sophisti-

144

cated mutualism. Paine notes that the subsystems are obviously important within food webs. A module often contains a set of resources and their consumers and can behave as a functional unit. Paine gives extensive references on strong interaction within modules in benthic and aquatic communities. It is clear that such modules are not simple structures. Coevolution has occurred, with predators often molding the community structure. The module is so highly coevolved that removal of a single species will cause the whole structure to collapse. The sort of coevolution that would lead to modules should occur under stable environmental conditions, so the most tightly organized systems may exist in relatively stable environments such as the tropics (Vermeij 1978).

Pimm (1980, 1982) has examined food webs from the ecological literature to determine if compartmentalization (or separation into modules) exists to a statistically significant degree. First, Pimm considered compartmentalization in the presence of different habitats, such as an arctic island that contains terrestrial, freshwater, and marine habitats (Summerhagen and Elton 1923). By comparing the measured food web with food webs created at random with the same connectance, Pimm and Lawton (1980) showed that compartmentalization indeed occurs in that case.

However, Pimm and Lawton (1980) did not find distinct modules within single habitats. Among likely candidates are two different plants and their insect species, or plants and their gall-forming parasitic species. In both cases, Pimm and Lawton found overlaps, among the insects on the two plants in the first case and in the enemies of the gall insects in the second case.

Pimm and Lawton (1978) did find that a partial separation into modules is common. Grazing and detrital chains are often partially separated. For example, in his studies of a salt marsh in Georgia, Marples (1966) found no overlap in primary consumers (herbivores and detritivores) but the predators in this system, spiders, fed on both of the food chains. Pimm and Lawton found that these distinct pathways to the primary consumer level (and often higher) occur in a wide range of terrestrial and aquatic systems.

In summary, the empirical studies indicate that food webs tend to contain an internal organization. This organization shows species grouped into subsystem or modules. These modules are not completely isolated from each other and may show some overlap. The result is a hierarchical structure that both analytical and simulation studies indicate should enhance stability.

SPATIAL SCALE AND HIERARCHY IN COMMUNITIES

We have reviewed evidence showing that species tend to aggregate into small, distinct subsystems. We now turn our attention to how these subsystems are arrayed in space. A problem of scale arises here since spatial aggregation may not be recognized if measurements are made on too large a spatial scale. To see how problems of spatial scale enter the picture, we must recognize that interactions among subsystems actually involve instantaneous meetings between individuals. However, measurement of the system involves observations over significantly larger time and space scales. The difference can be seen in the following example. Let us consider a system composed of a single food resource, X, a guild of three herbivores, H_i, and a single predator, P, that feeds on all three herbivores. We will assume that guild members compete very strongly with each other, so that the probability of any two of the herbivore populations being present in the same area is very small. Thus, if we look at the system at the scale at which the interactions occur (i.e., small spatial plots and short time intervals), there are three possible interaction matrices:

146

(1) Plots dominated by H_1

$$\begin{vmatrix} -a_{xx} & -a_{x1} & 0 & 0 & 0 \\ a_{1x} & -a_{11} & 0 & 0 & -a_{1p} \\ 0 & 0 & 0 & 0 & 0 \\ 0 & 0 & 0 & 0 & 0 \\ 0 & a_{p1} & 0 & 0 & -a_{pp} \end{vmatrix}$$

(2) Plots dominated by H_2

$$\begin{vmatrix} -a_{xx} & 0 & -a_{x2} & 0 & 0 \\ 0 & 0 & 0 & 0 & 0 \\ a_{2x} & 0 & -a_{22} & 0 & -a_{2p} \\ 0 & 0 & 0 & 0 & 0 \\ 0 & 0 & a_{p2} & 0 & -a_{pp} \end{vmatrix}$$

(3) Plots dominated by H_3

$$\begin{vmatrix} -a_{xx} & 0 & 0 & -a_{x3} & 0 \\ 0 & 0 & 0 & 0 & 0 \\ 0 & 0 & 0 & 0 & 0 \\ a_{3x} & 0 & 0 & -a_{33} & -a_{3p} \\ 0 & 0 & 0 & a_{p3} & -a_{pp} \end{vmatrix}$$

These matrices represent the three possible plots one would actually find in the field; that is, any sampled plot would contain only one of the three herbivores.

The key insight is that all three of these matrices contain a number of zeroes and as a result each can be decomposed into three by three matrices, each containing a single herbivore population. For the sake of argument, let us assume that the magnitude of the a's is such that all three of these matrices are stable. The conclusion would be that the system is stable.

However, this is not the way such food web matrices are usually assembled. Observations are made across a number of plots, and a's are averaged across space and time. The result is that the system is represented by a single matrix lumping together all species so that there are no zero values. Everything appears to be connected to everything else. In this case, we might discover that the system was unstable, and the conclusion would be drawn that this food web is not viable. As we saw above, this could well be the wrong conclusion.

This analysis leads us to the conclusion that analyses of food webs need to be scrutinized with respect to the scale at which the interactions are assumed to occur in the model versus the scale at which observations are taken. Representing these systems by a single matrix may obscure the fact that this system is hierarchically structured, with the guild H forming a holon that interacts as a unit with the resource and the predator. It would be interesting to find the conditions under which the three matrices (incorporating the hierarchical structure) were stable while the single matrix was unstable. This would represent the extent to which the hierarchical structure stabilizes the system.

A more sophisticated representation of the system would be required if more than one herbivore can be present, at least temporarily, on a plot. Under these circumstances, the zeroes in the three matrices would need to be modified. Going back to the logic of the model, an interaction between P and H is proportional to the probability of their meeting (HP) times a proportionality constant that expresses the conditional probability of an interaction occurring, given that they meet. On the plot where H_1 is dominant, the interactions between P and either H_2 or H_3 must be modified by the fact that P is less likely to meet individuals from these other populations on this plot. The probability of H_2's being on the plot dominated by H_1 is dependent on the interaction coefficient between H_1 and H_2. The larger the competitive interaction, the less likely it is to find H_2 on this plot. So we could calculate the conditional probability of P's interacting with H_2 on the H_1-dominated plot as another conditional probability, conditional now on the probability of finding H_2 on that plot.

One way to do this would be to assume that if two-way interactions between H_1 and H_2 were stable, then the two could coexist and the conditional probability of finding an individual of H_2 on a plot dominated by H_1 would be 1.0. In this case, we would use $a_{p2}PH_2$ to express the interaction of the predator in this plot. On the other hand, if the interaction coefficients were so large that the 2×2 matrix were unstable, then H_2 or H_1 would be eliminated and the probability of finding H_2 on the plot would be zero. At intermediate values of the interaction coefficients, we would assume a linear relationship. Based on the stability criterion for the 2×2 matrix ($a_{11}a_{22} - a_{12}a_{21} = 0$), we could write a new interaction term for P and H_2:

$$a_{p2}PH_2 \max (0, 1 - (a_{12}a_{21}/a_{11}a_{22})).$$

We could use this expression for recalculating the interaction coefficients between the predator and each of the herbivores on each of the plots.

This also suggests an approach to searching an n-dimensional alpha matrix for the underlying hierarchical structure. The algorithm involves something like: find the smallest unstable submatrix (say an interacting pair) and assume this is a "guild." Then split the full matrix into two matrices, each containing only one member of the guild, and repeat. The result would be a dendrograph of systems from which higher and higher levels of organization could be specified. It would be interesting to find large, highly connected (i.e., complex in the sense of Gardner and Ashby [1970]) systems that are really stable if they are hierarchically structured; that is, they only appear highly connected because individual, hierarchically structured exemplars of the system have been averaged across an inappropriate time–space scale to look like they are highly connected.

149

Another way to approach this question would be to use the matrix to devise individual systems (representing a particular plot) and perform Monte Carlo simulations of the system so that the average across the plots gives the interaction matrix. Then one could explore for structures that give stable plots but yield unstable systems when averaged across all the plots.

RESILIENCE AND CHARACTERISTIC TIME SCALES OF FOOD WEB PHENOMENA

Up to this point in the chapter we have discussed the structure of biotic communities and food webs. Because we have been operating within the population–community paradigm, it has been convenient to discuss hierarchical structure in terms of aggregates of tangible objects, that is, organisms and populations. It is necessary now for us to draw the connection between the tangible hierarchy of organism, population, guild, and community and the hierarchy of rates presented in Chapter 5.

We know from experience that a system of sufficient complexity can show many different phenomena with widely different time scales. As an example, consider the surface of a lake. Tiny transient ripples are governed by surface tension and minor disturbances such as insects, rising fish, raindrops, and so on. These waves have short wavelengths and high propagation frequency. Gravity waves are caused by wind roll across the lake with longer wavelength and lower frequencies. Finally, seiches are induced by steady winds along the length of the lake; they may oscillate back and forth with low frequency, say an oscillation per day. We intuitively recognize that the time scales of these phenomena are related to the spatial scales. The surface tension waves involve movement of only the top millimeter or so of water, the gravity waves involve the top meter or so, and the

150

seiches involve the whole body of water. Hence, there appears to be a hierarchy of dynamics at different time and spatial scales. This turns out to be true not only for a lake but for most kinds of systems.

When we think about what governs the multiplicity of temporal scales in a system, we recognize a dependence not only on system size but also on the number of components in the system. For example, two tuning forks with frequencies f_1 and f_2, that are very close, can together produce sound that has a much lower "beat" frequency, $f_3 = |f_1 - f_2|$. More complex patterns of beat frequencies can be produced when more tuning forks are added. So it appears that low-frequency phenomena can emerge from the interactions of components interacting at higher frequencies.

It is evident that this insight extends to biological systems. The activity within a cell seems like a frantic bustle to us, with rapid turnover of complex molecules. The turnover of cells is much slower, and the turnover of multicellular organisms much slower still. The application to biological and ecological theory of this well-known insight, however, demands some formalization of ecological time scales, which we develop below.

Up to this point we have been concerned with whether or not a community of species is stable in the Lyapunov sense, that is, whether or not the species return to the equilibrium following a perturbation. An equally important question is how quickly the populations return, given that the point is stable. It is easy to derive a mathematical expression for resilience for a single population described, say, by the Pearl–Verhulst equation,

$$dX(t)/dt = r[1 - X(t)/K]X(t). \qquad (7.6)$$

We introduce the new variable, $x(t)$, defined by $X(t) = K + x(t)$, where $x(t)$ is a small ($x(t) << K$) perturbation about the equilibrium point K. Substitution into Equation 7.6 yields, after dropping a term of second order,

$$dx(t)/dt = -rx(t) \qquad (7.7)$$

This equation can be solved to yield

$$x(t) = x_0 e^{-rt}, \qquad (7.8)$$

where x_0 is the size of the initial perturbation about K. The parameter r determines the time scale of the return to equilibrium. In particular, if we take as a measure of resilience the time, T_r, it takes for $x(t)$ to decay to $1/e$ of its perturbed value, then

$$T_r = 1/r. \qquad (7.9)$$

The value of T_r is the characteristic time scale of response of the system to a perturbation.

The situation for a large system of populations is more complex. There is not one but many time scales. These time scales correspond to the eigenvalues of the linearized system. Often, only one of these time scales, the longest, is of interest. We will talk more about this later.

Do populations, each with its own characteristic response time, interact to generate behavior on an entirely different time scale? DeAngelis (1975b) examined a pair of predator–prey equations where the zero isoclines have the configuration shown in Figure 7.7. This configuration results in limit cycle behavior, the amplitude and oscillation frequency of which can be estimated. DeAngelis showed the approximate period of oscillation, T, to be

$$T = 2\pi/(|K_{1,y}|K_{2,x})^{1/2}, \qquad (7.10)$$

where $K_{1,y}$ is the derivative of the growth rate of the prey X with respect to the predator Y, and $K_{2,x}$ is the derivative of the growth rate of Y with respect to X. Hence, a new time scale emerges that depends on the strengths of interactions

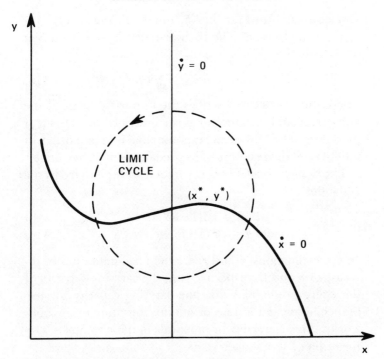

Fig. 7.7. Phase diagram for the predator–prey system studied by DeAngelis (1975b). The interaction results in a limit cycle with a time constant that may differ significantly from the population time constants.

between the predators and prey. The success of this approximation, however, depends on $K_{1,x}$ and $K_{2,y}$, the effects of X on its own growth rate and of Y on its own growth rate, respectively, being small compared with $K_{1,y}$ and $K_{2,x}$ near the equilibrium. Therefore, although the predator and prey interact to produce a new kind of behavior, it is not necessarily one that has a longer time scale.

Consider a model where explicit use is made of a hierarchy of time scales. Ludwig et al. (1978) modeled spruce budworm outbreaks in northern conifer forests. Three variables were involved, (1) budworm density, B, (2) the total surface area of the branches, S, and (3) the total energy reserve of the trees, E. The budworm density was described by the equation

$$dB/dt = r_0B(1 - B/K_b) - \beta B^2/(\alpha^2 + B^2) \quad (7.11a)$$

where the first term describes the intrinsic growth of the budworms and the second term describes the effect of predators. The budworm response time is on the order of months and the response of its predators somewhat longer.

The branch area and energy reserve were modeled by the equations

$$dS/dt = r_sS(1 - SK_e/(K_sE)) \quad (7.11b)$$
$$dE/dt = r_eE(1 - E/K_e) - pB/S \quad (7.11c)$$

In these equations, r_s and r_e are the maximum rates of increase of S and E, respectively, K_e the carrying capacity of the energy reserve, K_sE/K_e the carrying capacity for live branches, and p the rate of energy consumption by budworms. The characteristic time scale of these variables is on the order of seven to ten years.

It should now be added that the parameters in the budworm equation (7.11a) are not quite constants. The effects of predators will change as the amount of foliage, S, changes. In particular, increasing foliage, by providing cover for the budworms, decreases predator effectiveness and thus changes the parameter values in Equation 7.11a. Because S normally changes on a much slower time scale than B, one can assume that B is in equilibrium at any given time, and examine the slow change in this equilibrium value as a function of parameters by setting $dB/dt = 0$ and solving for B. Ludwig et al. (1978) found that after a slow increase in budworm density, a threshold would be reached at which that budworm equilibrium point literally disappeared. Biologically this was due to an escape from predation. The behavior of the equations at the threshold value becomes very complex. *B* increases drastically, causing a crash in *S* and *E* until *S* is small enough that the budworms are again controlled.

From the above example it appears that a natural hierarchy of time scales exists. Most of the time the system operates on a long time scale set by plant growth. But at critical moments, the fast time scale of insect reproduction takes over, setting the system back to a point at which the slower time scale again predominates.

The above phenomenon is quite common in nature. As K.E.F. Watt (1969) pointed out, this is the typical situation for infectious diseases. There is a long dormant phase in which the disease is endemic but kept to a low level by the relative scarcity of susceptible hosts. As host level increases on a slow time scale, the disease organism stays close to equilibrium until a certain threshold of host density is reached. Then the disease can spread on its own characteristic time scale.

Attempts have also been made to explain some of the long time-scale population cycles, such as the ten-year cycles of some tundra mammals. Oster and Takahashi (1974) showed that age-specific effects in populations can result in periodicities that have time scales exceeding the life spans of individuals in the population. Finerty (1980) reviewed the possibilities of beat phenomena between intrinsic population periodicities and extrinsic periodicities but found no conclusive evidence of this occurring.

The existence of fast and slow time scales in a variety of ecological contexts, as shown above, suggests that there might be a general hierarchical ordering of time scales across a wide range. As we will discuss in Chapter 9, the dynamics of the fastest variables may be described by equations in which slower variables appear as parameters. The changes in these parameters are in turn described by equations in which still slower variables appear as parameters, and so on.

Now we wish to inquire into the question of whether the existence of relatively weakly coupled modules within a community is likely to give rise to phenomena on new and longer time scales. Consider a system linearized about an equilibrium point, described by Equation 7.2. If the community is structured into weakly interacting modules, then A has the form:

$$A = \begin{vmatrix} A_1 & B_1 \\ B_2 & A_2 \end{vmatrix}$$

where A_1 is a $p \times p$ submatrix and A_2 is a $q \times q$ submatrix. The elements in the diagonal blocks, A_1 and A_2, are of zero order, while those in the off-diagonal blocks, B_1 and B_2, are of first order, hence, relatively small. The eigenvalue equation for this matrix is

$$[\lambda^p + b_{1,p-1}\lambda^{p-1} + \ldots + b_{1,0}]$$
$$[\lambda^q + b_{2,p-1}\lambda^{p-1} + \ldots + b_{2,0}]$$
$$+ c_{p-2}\lambda^{p-2} + c_{p-3}\lambda^{p-3} + \ldots + c_0 = 0. \quad (7.12)$$

where all terms c_i are of the first order or smaller. Note that the two brackets are the characteristic equations of submatrices A_1 and A_2 respectively, with $b_{1,0} = \det(A_1)$ and $b_{2,0} = \det(A_2)$. Unless $b_{1,0} = 0$ or $b_{2,0} = 0$, it can be shown that the terms with coefficients c_i are only first-order perturbations on the eigenvalues determined with the c_is set to zero. If $b_{1,0} = 0$ or $b_{2,0} = 0$, at least one eigenvalue occurs that is determined in part by the c_i values. This is entirely equivalent to what occurred in the simple predator–prey model discussed above, where a limit-cycle frequency governed by the interaction strengths was generated.

If species modules and not the interactions among these modules largely determine the characteristic time scales (e.g., of response to a perturbation) of a community, then it is important to look closely at what governs the frequencies of each module. In particular, are there structural features of a module that affect the emergent frequencies of the whole system? Pimm and Lawton have explored a number of themes related to this general question. Of particular interest are their views on what limits food-chain length to four or five trophic levels. Pimm and Lawton (1977) advocate the view that decreasing resilience of the system, as trophic levels are added, limits food chain length. This view contradicts the more widely held view that decreasing available energy to the top level is the limiting factor. The views of Pimm and Lawton seem to be supported by evidence that world ecosystems differ in primary production by up to four orders of magnitude, for example, between tropics and tundra, while there is no evidence of chains in the productive areas being appreciably longer than those in unproductive areas.

Pimm and Lawton (1977) based their argument on analysis of the Lotka–Volterra equations linearized about an equilibrium. They calculated the return time of the system according to

$$T_r = 1/[\text{real} (\lambda)]_{\text{max}}.$$

They found that as the chain length increased, T_r also increased. Thus, it appears once again that the slower response times, which are associated with higher levels of hierarchical organization, are a natural consequence of the way species populations are organized into food webs.

CONCLUSIONS

Our consideration of the ecosystem as a network of species has yielded considerable theoretical evidence, and at least some empirical evidence, that the system is hierarchically structured. We have shown that the intuitive idea of hierarchical structure's stabilizing the system can be given more precise meaning in the context of communities. Most importantly, we have tried to show that hierarchy theory can be taken a step further and used to develop hypotheses about how natural food webs might be assembled.

CHAPTER 8

Ecosystems as Hierarchies of Processes

The last chapter outlined the implications of hierarchy theory for populations, communities, and food webs. Hierarchy theory is extended in this chapter to the analysis of functional phenomena, that is, the processing of energy and matter. The objective is to discover a general concept of the ecosystem that will be consistent with the known facts about the natural world and that will avoid the pitfalls and limitations pointed out in previous chapters. We begin by examining the perspective that hierarchy theory gives us on ecosystem processes.

CONSEQUENCES OF IGNORING HIERARCHICAL STRUCTURE

The natural world as a processor of energy and matter is extended over a wide range of spatiotemporal scales. When one approaches the natural world with a particular scientific question in mind, one focuses on a specific spatiotemporal scale of observation, for example, a forest stand. One can see the components of this system, its primary producers and decomposers, for instance. One can see some aspects of the larger system of which the forest stand is a part. These larger aspects appear as constraints or boundary conditions and are considered as the context or environment of the ecosystem.

Such a conceptualization is in keeping with the way ecologists have studied ecosystem processes. The implied hierarchical levels are familiar. The problems arise in extrapolating from this specific observation set to other scales of observation. It is tempting to talk about the ecosystem as if one specific observation set were optimal or absolute. In truth, the specific observation window chosen and the specific phenomena under investigation are set by the purpose of the study. Change the purpose, and the system of interest changes, for example, in spatial and temporal extent. Thus, in an input–output analysis of water-soluble nutrients, the watershed becomes the unit of study and the forest stand now becomes a candidate for a component of the system. The ecosystem comes to look more and more like an abstraction with spatiotemporal properties that can be specified only within the context of an observation set. The ecosystem as an independent, discrete entity begins to become less tenable.

Initially, the insight that ecosystems are hierarchically structured seems just a matter of common sense. The investigator must be aware of the spatial and temporal scales on which phenomena of interest occur and must take this scale into account in designing experiments. Therefore, one could argue that discussions of scale and hierarchy are interesting but good investigators already know about them. The principles outlined in this book would therefore likely be true, but only in a methodological sense and we would be preaching to the converted. To show the fallacy of this argument requires that we examine ways in which explicit consideration of scale provides counterintuitive results.

Ignoring Scale in Evolutionary Studies

Gingerich (1983) discusses an excellent example of the errors that result when scale is ignored. He discusses the measurement and interpretation of rates of evolution. The rates are measured as morphological changes expressed in darwins (change by a factor of e per million years, where e is the base of the natural logarithms). In laboratory selection experiments, observations are made at intervals of 1.5 to 10.0 years. Many discrete changes can be observed and rates average about 60,000 darwins. In colonization studies, fewer observations are made over longer intervals (70–300 years), fewer changes are scored, and measured rates average about 400 darwins. Changes in postpleistocene mammals involve observations across intervals of 1,000 to 10,000 years, and morphological changes are scored at a rate of about 4 darwins. Finally, in fossil assemblages measured across millions of years, very few of the minor changes can be observed at all and this omission results in measured rates of about 0.1 darwins.

By regressing 521 separate measurements against the time interval over which the observations were made, Gingerich measures the systematic bias that has been introduced by ignoring scale. He reviews the literature that addresses why the rates of evolution of postpleistocene mammals are an order of magnitude faster than older assemblages of invertebrates. He then shows that this has been a false argument since, once the data are corrected for the scale bias, the invertebrates may have changed several times faster than the mammals! In other cases, discussions of differences in evolutionary rate have been misguided since the corrected rates show no significant difference at all. This analysis demonstrates that errors can creep into scientific investigations if problems of scale are not explicitly considered.

Ignoring Constraints in Ecosystem Models

The dynamics of an ecological entity, such as a population, are constrained by the higher levels of the hierarchical system of which it is a part. The same population incorporated into a different system may experience a different set of constraints. If we observe the same population across a number of systems, we find that it has the potential for a wider range of behavior than it actually displays in any single context. For example, we can observe a far more rapid rate of reproduction of a population in the laboratory than we can ever actually measure under field conditions.

Ignoring such hierarchical constraints may explain why some ecosystem models perform poorly. It is clear that both the potential and the constrained dynamics of ecosystem components must be considered if we are to understand ecosystem dynamics. Yet it appears that some ecosystem models emphasize one almost to the exclusion of the other.

In this context, linear ecosystem models can be viewed as overemphasizing constrained behavior. It has been argued (Patten 1975) that since ecosystems actually behave as though they were linear, linear models are an appropriate approach to ecosystem dynamics. For example, ecosystems can be observed to return asymptotically to some nominal operating state following minor perturbations. This type of behavior is typical of a linear system. The variety of behavior typical of nonlinear systems, such as limit cycles, is seldom observed under normal conditions.

The basic limitation of the linear model is that it emphasizes what the system is allowed to do under the current, undisturbed hierarchical constraints. The restrictions on the applicability of the linear model (O'Neill 1979b) are precisely that the situation being simulated does not involve any perturbation likely to disrupt the hierarchical structure. But, as a result, the linear model cannot simulate what the system is capable of doing if the constraints are changed. While the model cannot respond unstably, the ecosystem can. Basically, the linear model emphasizes what the ecosystem is allowed to do within established constraints and ignores what the ecosystem is capable of doing if these constraints are lifted.

On the other hand, many nonlinear models of ecosystems

overemphasize what system components are capable of doing. The models represent the range of possible behaviors, but few are capable of constraining the behaviors to the restricted set actually observed. The reason that nonlinear models can represent what components are capable of doing is that they often incorporate relationships and parameter values measured in independent, controlled experiments. The parameters express how rapidly a component can change under ideal laboratory conditions, rather than how rapidly it will change under constrained field conditions. The models typically have erratic behavior that is never observed in any actual (and therefore constrained) ecosystem. The models represent what the system can do instead of what it will do.

T.F.H. Allen and Starr (1982) pointed out that the basic problem with many ecosystem models is that they do not account for the hierarchical constraints that operate in the ecosystem. Instead, the tendency is to deal with the lowest-level entities in the system. For example, in an observation set that permits identification of individual populations, each population is regarded as a separate state variable. Then, each population is represented as interacting directly with every other population. In fact, further analysis would reveal that the individual populations form guilds, trophic levels, and other organizational levels. If the models were constructed to account for all of the levels of organization, then more of the constrained behavior of the system would be accounted for. The natural constraint structure could be built into the model and the model could explain what actually happens as well as what can happen.

THE CONCEPT OF INCORPORATION

No specific spatiotemporal scale encompasses the ecosystem and, therefore, no specific level in the system can respond stably to all scales of perturbation. However, it remains a fundamental insight that at each scale the system is homeorhetic (i.e., returns to preperturbation dynamics) and self-organized to some degree. At the level of the forest stand, for example, the ecosystem retains and recycles nutrients. This frees primary production from the limitations imposed by erratic inputs of nutrients from outside the system.

The question now becomes whether conceptualizing the ecosystem as having levels of organization, each of which is homeorhetic relative to a specific scale of perturbation, gains us new insight. Such insight may result from considering how levels in the hierarchy respond to perturbations on a larger and larger scale. In many observation sets, the ecosystem appears to be a nested hierarchy in which the higher levels subsume and contain the lower levels. This nestedness is the key to understanding how higher levels avoid the catastrophic consequences of perturbation. What one observes is a phenomenon we call incorporation.

A perturbation comes from *outside* the system; it is, by definition, uncontrolled and capricious when viewed by components of the system. Consider hourly carbon fixation by trees broadly scattered across an open field. The rate of fixation is dependent upon, and therefore constrained by, factors such as oscillations in temperature and wind. The short-term fluctuations appear as capricious, outside the control of the trees.

Let us now change the observation set to trees closely grouped into a forest stand. The stand now dampens temperature fluctuations and slows air movement so that carbon fixation by the individual tree is no longer affected by short-term fluctuations in environmental variables. As early as 1922, Larsen noted that air temperatures under a closed canopy were 10° C warmer at night and 10° C cooler by day than the same area following clear-cutting.

By grouping the trees, a new organization level, the forest, has appeared and the perturbation has been incorporated. We can legitimately speak of incorporation whenever we see ecosystem organization exerting control over some aspect of the abiotic environment that is uncontrolled at a lower level of organization. We can see this phenomenon occur across all of the spatiotemporal scales that characterize the ecological system. For example, precipitation must certainly be considered as an external, uncontrollable perturbation. However, considered at the scale of an extensive vegetated landscape, even this aspect of the environment can be incorporated.

Over extensive forests, evapotranspiration provides a large fraction of the moisture of the overlying air mass. As a result, the presence of vegetation encourages rainfall. The importance of the interaction becomes patent when the vegetation is removed, for example, by overgrazing. Lowered evapotranspiration leads to less rainfall and further stress on the remaining vegetation. This process of desertification has turned many areas (e.g., the Middle East) from continuous vegetation to desert within historic times.

In some cases, perturbations are incorporated by temporal mechanisms. An excellent example is nutrient cycling. Without the functional components that retain and recycle nutrients, the rate of production would be constrained by the rates at which nutrients arrive from outside the system. One also thinks of dominant organisms that actively incorporate perturbations. For example, alligators build ponds that form refuges for aquatic species that could not otherwise exist because of seasonal fluctuations in water level.

In other cases, perturbations are passively incorporated in a spatial framework. This type of incorporation is discussed by Shugart and West (1981). Only systems that are large relative to their perturbations maintain a relatively constant structure. Figure 8.1 is adapted from Shugart and

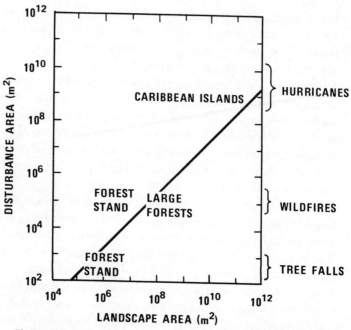

Fig. 8.1. Relative size of disturbance and landscape units. Above the diagonal line are disequilibrium systems, such as Caribbean islands, that are the same size or smaller than their characteristic perturbations, such as hurricanes. Below the line are more constant systems that are large relative to their perturbations. Redrawn from Shugart and West (1981).

West (1981) and shows the average size of perturbations affecting a variety of ecosystems. Carribean islands are small compared to hurricanes and cannot avoid being constrained by these storms. On the other hand, Appalachian forests are large compared to the average wildfire and are relatively constant in composition over time.

A number of authors have also noted that community structure seems to incorporate large-scale perturbations. Mutch (1970) pointed out that fire is the factor that ensures the continuance of the fire-adapted plant community. This community exists because of fire, not in spite of it. Waring and Franklin (1979) have suggested that massive, long-lived coniferous trees represent an adaptation to the regional climate of the northwestern United States. Without considering the broad spatial scale that characterizes this regional climate (i.e., by focusing on individual stands) the massive structure of these forests appears to be an anomaly. At an even broader scale, Davis (1976) argued that the glaciation history of North America has altered life histories of tree species, allowing them to penetrate closed canopies following glacial retreat.

Ecosystem Incorporation of Fire

Some of the best examples of ecosystem incorporation have been documented in fire ecology. Mooney et al. (1981) provide a summary which emphasizes ecosystem properties. It is clear from their discussions that extensive forests in western United States have passively incorporated forest fires. Because the system is distributed across a broad and diverse landscape, some portions are not destroyed by a fire, which permits recovery of the entire system. The system is very large relative to the size of the fire. The perturbation is incorporated in the sense that the survival of the forest is not threatened by fire.

Spatial incorporation can ordinarily be restated in terms of frequencies and constraints. Zedler et al. (1983) present an interesting case for the California chapparral. This fire-adapted community recovers rapidly from small fires. However, if a second fire recurs on the same area within a single year, this exceeds the recovery capability of the system and drastically alters the vegetation. In this case it is clearly fire frequency that causes the effect rather than the presence of fire itself.

It is also illuminating to view fire incorporation from the perspective of observation sets. Let us limit the extent of our observation set to the dimensions of a single small stand

167

within the forest. At this scale, we see periodic fire destroying the system that slowly grows back to its original biomass. The slowest frequency component of the data set would be the frequency of the fire. The fire would appear as external to the system and as capricious and uncontrollable. At this level of resolution, fire is a perturbation.

Let us now extend the spatial extent of the observation set to include refuges and sections of the landscape that had previously burned and now were serving as seed sources for pioneer species. Biomass, integrated across this landscape, now changes slowly relative to the recurrence frequency of the fire. At this level of resolution, the system appears to have incorporated fire. Thus, fire is and is not a perturbation depending on the scale of observation. T.H.F. Allen and Starr (1982) report multivariate analysis of a prairie where a similar change in observation sets shows fire as having these dual characteristics.

Of course, we can also find examples in which the ecosystem incorporates fire by a combination of temporal and spatial mechanisms. The chapparral (Allen and Starr 1982) is extended in space so the probability of a fire perturbing the entire system is small. But, in addition, the system changes fire frequency by production of volatile oils which encourage more frequent fires. The more frequent the fire, the smaller the accumulated fuel in the form of detritus and the smaller the fire will be. By increasing the frequency of the fires, the spatial extent needed to incorporate fire becomes less.

A number of authors have pointed out that fire-adapted species seem to manipulate fire frequency. Mount (1964) argues that eucalypts control fires with low moisture content and high oil content in the leaves. The leaves are relatively flammable and tend to encourage small, frequent fires. L. J. Webb (1968) points out that fires are encouraged in eucalypts by explosive gases composed of eucalypt oils.

168

Mutch (1970) indicates the general flammability of litter in other fire-adapted species such as ponderosa pine and tropical hardwoods. In many systems, fire-adapted species will disappear unless fire occurs frequently (Noble and Slatyer 1980). If fires are infrequent, longer-lived species tend to take over. Gill (1975) reviews the hypothesis that vegetation actively manipulates fire frequency. He concludes that the hypothesis seems reasonable but has yet to be subjected to experimental testing.

Summary

Viewed at the proper level of resolution, perturbations may often be incorporated. The perturbation is no longer external to the system, no longer uncontrollable. The resulting system appears to be more stable both in the sense of more persistent in time and also in the sense of more constant in composition. Thus, incorporation can be proposed as a mechanism whereby ecosystems are stabilized.

The phenomenon of incorporation seems to be less often observed in aquatic systems simply because the environment is less stressful and less variable. We can, of course, find cases where kelp, coral, or macrophytes modify the effects of wave motion. Here it is the amplitude or severity of the perturbation that is modified. We may also point out that oscillations in nutrient supply are incorporated by nutrient cycling just as they are in terrestrial ecosystems.

In both terrestrial and aquatic systems, incorporation provides an organizational principle that links the system across hierarchical levels. The system viewed at one spatiotemporal scale is homeorhetic to perturbations at that scale. The ecosystem considered on this scale will be unstable to larger, slower perturbations. However, by incorporation, these larger pertubations will become part of the ecosystem by being brought under feedback control at a larger scale of organization. Thus, as one moves up organizational levels, perturbations are incorporated and homeorhesis is established over a broad range of perturbations.

The phenomenon of incorporation also helps explain why it is difficult to account for ecosystem phenomena in terms of species populations. At higher levels of organization, many aspects of the abiotic environment have been incorporated into the structure of the system. Thus, soil becomes part of the system in nutrient cycling. In a very real sense, the abiotic components become a part of the ecosystem and cannot be considered as external (Webster 1979). Thus, the definition of the ecosystem as plants and animals and their environment is inadequate. The ecosystem is composed of plants, animals, incorporated abiotic components, and the environment. At large scales, ecosystem function will seldom be explicable in terms of biotic interactions alone.

INSTABILITY IN A HIERARCHICAL SYSTEM

The observation that functions, such as nutrient cycling, continue even though individuals and populations appear and disappear has provided strong motivation for the study of ecosystems. One of the most important ecosystem questions deals with the level of perturbation that will cause the system to become unstable. Therefore, it is natural that we should consider what hierarchy theory can tell us about a system when it goes unstable and establishes a new organizational state.

Conceptually, what happens is that the system of constraints is disrupted and the components of the system begin to operate as rapidly as they are capable of acting. Once the constraints are lost, the hierarchical organization is lost. When a system goes unstable, it is the normal functioning of the unconstrained components that tears the system apart.

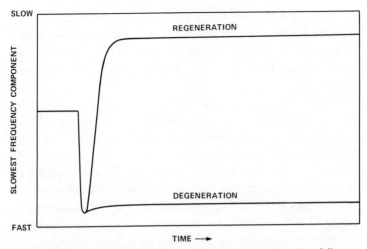

Fig. 8.2. Response of a system to a drastic perturbation. At the point of disturbance, the organization is disrupted and the slow frequencies associated with higher organizational levels disappear. After regeneration of system organization, these slow components appear once again. If recovery does not occur, the system degenerates and the slow-frequency components do not reappear.

What happens next will depend on the new set of constraints it finds.

The system can regenerate a new organizational structure or it can degenerate (T.F.H. Allen and Starr 1982). The difference between the two can be seen in Figure 8.2, which shows the lowest frequency that can be observed during the transitions. At the point of instability the low-frequency components disappear and we observe only rapid, erratic signals. Eventually, a new set of constraints will form and a new hierarchical structure can be detected. If the new organizational state shows frequency signals that are as low or lower than the system before the instability, we say that the system has regenerated a new organization. If the new organization shows only relatively high frequencies, then we say that the system has degenerated because it is now the

171

victim of a wide range of environmental perturbations. It cannot dampen out oscillations that are of lower frequency than its own slowest rate (see Chapter 5). It no longer has the organizational levels needed to incorporate perturbations.

Notice that recovery to exactly the same organizational state would merely be a special case of the process. If the perturbations were small, the original constraint structure would not be destroyed and we would expect recovery. If the perturbation completely ruptures the constraints, then the original structure is lost and there would be only a small probability of returning to exactly the same organizational state. Thus, a return to the same state is a likely consequence of a small perturbation but an unlikely consequence if the perturbation is severe.

It is helpful to view the regenerated and degenerated systems in terms of the observation that a low-frequency component can ignore high-frequency signals, but not vice versa. Consider a simple input–output experiment in which we subject the system to perturbations of various frequencies (Fig. 8.3). The degenerated system is affected by perturbations over a broad range of frequencies. The regenerated system can ignore changes at many of these frequencies.

Examples of degenerated ecosystems are rather easily found. Adjacent to smelters in Sudbury, Ontario, and Copper Hill, Tennessee, severe pollution disrupted ecosystem organization. The system has degenerated to bare, eroded ground sparsely covered by low vegetation. It maintains a low level of organizational structure since it maintains homeorhesis only over a narrow range of perturbations.

A forest fire also clearly precipitates a degeneration relative to the local forest stand. Immediately following the fire, the system is essentially dead and passively responds to any environmental signal. Of course, if we change our observa-

172

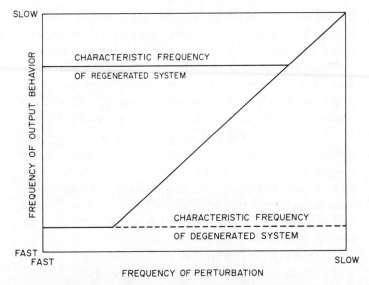

Fig. 8.3. Response of a system to different frequencies of disturbance. The regenerated system has slow components associated with higher organizational levels. Therefore, system behavior remains relatively constant in the face of short-term and rapid oscillations in the environment. The degenerated system has no slow components and shows a high characteristic frequency. Therefore, it is unable to dampen oscillations in the environment that are slower than this characteristic frequency. As a result, its output behavior (diagonal line) is directly affected by low-frequency perturbations.

tion set to include a large enough area or a long enough time, we can see that the forest stand eventually recovers to something near its original organizational structure.

POSITIVE FEEDBACK WITH CONSTRAINTS IN ECOLOGICAL SYSTEMS

We have become accustomed in ecosystem ecology to adopting engineering approaches to ecological problems. Because of the importance of negative feedback as a control device in engineering systems, it has been broadly suggested as a mechanism controlling natural systems. It is this type of control that is suggested in many population models (Chapter 7). These models often contain a negative term that is a function of the square of the population. The constraint is represented as a self-constraint, a negative feedback of the population size upon its own growth.

We are certainly not the first to question the relevance of negative feedback as a palliative for all ecological control. Waide et al. (1974) offered a number of reasons to question the applicability of classical engineering techniques in ecosystems (see Chapter 3). DeAngelis et al. (1986) have provided extensive evidence for the presence of positive feedback in ecological systems. Positive feedback as a common and important factor in the formation of system structure has also been discussed by Maruyama (1963).

Examples of positive feedback with constraint are particularly easily found at large spatiotemporal scales. Indeed, they may be characteristic of global environmental processes (Botkin, in press). In a previous section we discussed the positive feedback between devegetation and rainfall which can result in desertification. Similarly, glaciation probably begins with a slight lowering of temperature. The glacier begins to grow and change the albedo of the Earth's surface. This further lowers the temperature and encourages further glacial growth. The positive cycle proceeds until the system meets a new constraint, probably the lowering of the oceans and the reduction in the supply of water available for further expansion of the glacier.

Saltatorial Evolution

One consequence of the renewed interest in positive feedback is the proposal that saltatorial evolution is to be expected in hierarchical systems (Knoll et al., in press). For much of the time, novelties introduced by genetic change will be repressed by natural selection. Under this constraint system, change will be minor and slow.

174

The situation changes radically when a minor additional genetic change succeeds in breaking through a constraint. An example of such a change was the alteration in metabolic pathways which permitted diatoms to develop a silicon shell. The shell protected them from predation. As a result, there was an explosive radiation of diatoms that continued until a new constraint was reached. The constraint was simply that the concentration of silica in the ocean waters dropped to a level where it was too difficult to extract.

Once a genetic novelty succeeds in breaking a constraint, we would expect rapid change in a hierarchical system. Thus, the observed fossil record of long periods of slow, stepwise change, punctuated by brief periods of explosive adaptive radiation is precisely what would be expected in a hierarchical system. No novel mechanisms are needed to explain saltatorial evolution, which is simply a consequence of the way the system is organized. In fact, no drastic alterations in the environment, such as radical changes in temperature, are required. Even without radical jumps in environmental conditions, simple mutations can produce the observed response.

IDENTIFYING THE COMPONENTS IN ECOSYSTEM PROCESSES

Throughout our presentation, we have maintained that arbitrary designation of system components was a dangerous proposition. Traditional spatiotemporal scales (see Chapter 2) tend to limit the components to tangible objects, perceivable at human levels. Reliance on the traditional concept of levels (Chapter 4) leads one to assume that ecosystem dynamics are always explicable in terms of population interactions.

We must now apply hierarchy theory to finding the relevant components of ecosystem function. To study a specific

175

phenomenon, one chooses an observation set at the space and time scales that characterize that phenomenon. To study ecosystem processing of soluble nutrients, one measures precipitation inputs and streamflow outputs on the scale of a watershed. The objective is the simplest set of components that will explain the relationship between the nutrient inputs and the nutrient outputs. We will say that the phenomenon or process has been "explained" if the phenomenon can be shown to be the consequence of interactions among the components.

The direct approach to this problem is to analyze observations on the system and allow the data to force upon us the relevant components of explanation. We will use the available information on the system to extract the minimal set of components. This approach is common to many fields and falls under the general category of "systems identification" (see, e.g., Walter 1982).

Consider the dynamics of an element such as calcium or magnesium in an organism. A simple experiment would be to feed the animal on food labeled with a radioactive isotope of the element. Feeding would continue until the level of isotope in the animal's body no longer inceased with time. At this point, the isotope is withheld and the decreasing body burden is measured. This output data is analyzed to identify dynamic components and the rates at which these components process the element. Ordinarily the data resemble a decreasing exponential function and the body burden through time, $X(t)$, is described by:

$$X(t) = \sum_{i=1}^{n} a_i \exp\left(-b_i t\right) \qquad (8.1)$$

where a and b are constants extracted by nonlinear least squares. Typically, as many as three terms for Equation 8.1 (i.e., $n = 3$) can be extracted statistically from data on $X(t)$ over time.

Actually, the form of Equation 8.1 was chosen not only because the data resemble an exponential curve but also because the equation is the solution to a simple model describing movement of the isotope through the animal's body. The simplest version of this model would consider three compartments lined up in parallel. The b_i represent the rate at which the element is turned over in each compartment and the a_i represent the fraction of the total body burden found in each component at the beginning of the elimination portion of the experiment.

Although one ordinarily would need more complex models to describe dynamics, this example serves our present purposes. The example shows that it is possible to make a set of observations on the system (measurements of total body burden over time) and to work directly with this observation set to derive the components that explain the data. It was not necessary to make any a priori assumption about what the components would be. There was no need to force the phenomena into a set of components that are observable and interesting at some other space and time scale.

This approach has been widely used in watershed hydrology to develop models relating precipitation input to streamflow output. This would seem to establish the applicability of the approach to many watershed scale studies of nutrient retention and cycling. The approach has also been used to develop models of aquatic ecosystems (Roberts and DiCesare 1982) that explain primary production and nutrient dynamics. In general, the approach is applicable whenever one is relating inputs to outputs or when one is analyzing integrated responses of the ecosystem.

Applicability of Systems Identification

It appears, then, that it is possible to define objectively the functional components of an ecosystem, at least under some

177

experimental conditions. However, an objection immediately suggests itself: How common is it to measure the input–output behavior of a system? Does this approach represent an unreasonable limitation of ecosystem studies to the measurement of total system response? It appears to be far more common to perform studies on some limited aspect of the system, for example, some aspect of remineralization of nutrients in the soil. Does the hierarchical approach imply that such studies are without merit?

In fact, the use of total-system parameters is more common than appears at first. Many studies of nutrient cycling depend on precipitation input and streamflow or lysimeter output. Aquatic studies, both in lakes and streams, have used chemical parameters measured in the water column (CO_2, P, N) as integrated indicators of system performance. Much of the experimental work on laboratory microcosms, considered as surrogate ecosystems, has used total-system parameters (Van Voris et al. 1980; Waide et al 1980). The list of total-system indicators that might be usefully applied to ecosystem studies is extensive. Many suggestions are offered by Hammons (1981).

However, in most studies one would not rely exclusively on total-system measurements. Rather, one would combine the integrated responses of the system with more detailed measures of individual processes. The problem with focusing exclusively on individual-process studies is that they may make unwarranted assumptions about system components. If no information is available on overall system behavior, then it is difficult to establish that the tangible components under study represent the needed functional components. Instead of measuring ecosystem behavior such as energy or elemental processing at the spatiotemporal scale at which it occurs, the process study may measure tangible objects at a different, perhaps more convenient scale, and the attempt to assemble the behavior of the overall system is likely to fail.

The real question that must be answered is: How do the functional components identified by systems identification techniques correspond to the tangible objects we can identify and study at other scales of resolution? We must now turn to this important question.

How Tangible Are Functional Components?

In general, the problem of matching overall ecosystem behavior with components measurable at another level of resolution is a difficult one. In an earlier chapter (Chapter 4) we discussed a study of Rosen (1977a) in which he demonstrated that one could not in general reconstruct system behavior knowing only component behavior. When Rosen's system was divided into functional components these did not unambiguously correspond to tangible components.

We can illustrate the problem with output from a ten-component model described by Equation 8.1 (O'Neill 1979a). Consider "data" from this model corresponding to daily measurements with a random error of $\pm$ 2 percent. The original generating equation could be fit to these "data" with $R^2 = 0.985$. If the data are fitted to Equation 8.1, with no foreknowledge of the number of components in the model, one obtains:

$$X(t) = 499 \exp(-0.009t) + 56 \exp(-0.215t) \quad (8.2)$$

with an $R^2 = 0.984$. Thus, the two-component model explained almost all of the variance in the "data." However, only two components are needed and the coefficients of Equation 8.2 do not correspond in any simple way to the coefficients of the original model. The functional components in Equation 8.2 do not simply reduce to the "tangible" components of the original equation.

It is important to recognize that this example is not pathological. It is common to find that measured turnover rates do not correspond to the assumed components of the system (Lassen and Henriksen 1983). In isotope feeding ex-

179

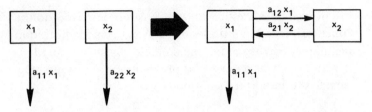

Fig. 8.4. Simple two-component systems. The system on the left has dynamic behavior that can be immediately identified with the turnover rates of its components. The system on the right is more complex and its dynamics do not correspond simply to the turnover rates of its components considered in isolation.

periments that identify three functional components, it is usually assumed that the fastest component represents gut turnover rate, the slowest perhaps turnover of bone matrix, and the intermediate coefficient is soft body tissue. However, any attempt to measure directly bone turnover rates fails to come up with the same value.

A similar situation has been experienced by hydrologists. In matching inputs and outputs from a watershed, they identify three turnover rates. The fast one is identified with precipitation falling directly on the stream channel. The slow one is associated wih seepage into ground water and very slow movement toward the stream. The intermediate turnover rate is assigned to a poorly understood process known as "interflow" that is supposed to represent movement through upper soil layers. However, once again, it is essentially impossible to match measurements of horizontal movement of water in superficial soil layers with the turnover rate needed to explain overall system output.

A numerical example will help explain what is occurring. Consider the system shown on the left side of Figure 8.4. Here we have a two-component system, with each component losing nutrients at a characteristic rate, a_{ii}. These are the loss rates we would measure if we isolated the components and studied each separately. If we set $a_{11} = 0.11$ and

$a_{22} = 0.10$, then the loss of nutrients from the system $(X_1 + X_2)$ per unit time would be governed by

$$dX/dt = AX \qquad (8.3)$$

where X is the $n = 2$ vector of components and A is a 2×2 matrix of exchange coefficients:

$$A = \begin{vmatrix} -0.11 & 0 \\ 0 & -0.10 \end{vmatrix}$$

The turnover rates are determined by the eigenvalues of the matrix A. The eigenvalues are the solutions to the equation:

$$s^2 - (a_{11} + a_{22})s + (a_{11}a_{22} - a_{12}a_{21}) = 0$$

Substituting the values of the matrix into the equation shows that the eigenvalues are -0.10 and -0.11. Thus, the system turnover coefficients are directly associated with the component turnover coefficients. In this case, there is no problem with reconstructing total system behavior knowing only the isolated properties of the components.

Let us now consider the system shown on the right side of Figure 8.4. Here, a significant portion of the nutrient loss from X_1 is taken up by X_2. In turn, X_1 is able to take up all of the nutrient released from X_2. In other words, the system corresponds to a simple nutrient cycle. For the sake of comparison, we will keep the turnover rates of X_1 and X_2 at 0.11 and 0.10 as in the previous example. But now, because exchanges occur between the components, the matrix A in Equation 8.3 becomes:

$$A = \begin{vmatrix} -0.11 & 0.10 \\ 0.10 & -0.10 \end{vmatrix}$$

Once again we can substitute these values into the equation for the eigenvalues to find the rates which characterize this new system. The eigenvalues now turn out to be -0.205 and -0.0049! Thus, the characteristic rates of this new system turn out to be poorly characterized by the turnover rates of the two components. If we were to measure the output of the system, we would discover the eigenvalues, not the coefficients of the individual components.

Thus, in general, one cannot reassemble the behavior of the system knowing only the behavior of the isolated parts. It is possible to reconstruct system behavior if the components are measured within the system, but then measurements are needed of the interactions as well as the turnover behavior of the components. The general problem of what measurements will ensure that the components and all parameters can be properly identified is treated in detail by Walter (1982). It will suffice for present purposes to point out that measurement of total system behavior with simultaneous measurement of a few selected points within the system (e.g., soil water or nutrients fixed in new leaf tissue for a terrestrial nutrient cycling problem) is probably the best strategy.

The lesson is that the functional components needed to explain ecosystem behavior will not always be immediately apparent. Even if the components correspond to tangible entities, one may still need to measure the total system to reconstruct all of the component interactions. Thus, although the behavior of the ecosystem is nothing other than the behaviors of the tangible, interacting components, still the behavior of the ecosystem needs to be addressed and measured at the spatiotemporal scale at which it occurs.

Applications of the Functional Approach

We would now like to illustrate how the approach advocated in the last few sections can be applied. We will do this by reviewing some published studies that have followed the approach by taking measurements on the intact system, allowing the measured time series to determine the functional components of the system.

Emanuel et al. (1978) examined the output of FORET, a model for forest dynamics. They transformed the time series into a frequency spectrum and discovered that the system had higher levels of organization that were represented by low-frequency signals. Allen and Shugart (reported in T.F.H. Allen and Starr 1982) applied multivariate data reduction and proposed a hierarchical control system to account for FORET's behavior. Dominant trees prevail for several hundred years. Throughout that period, the dominant tree is the context for all other growth in the plot and higher frequencies related to browsing and climate have little effect. These latter factors become significant when the old tree dies and a successor appears. The responses of competitors for stand dominance is determined by probabilities that derive from range and distribution data for the various species. Thus, the hierarchical levels exhibited by the model derive from the interplay between three constraints operating at different time scales: recent stand history giving the dominant tree, climatic and biotic constraints upon recruitment, and the probabilities imposed by geographic species-distribution data.

Dwyer and Perez (1983) monitored large marine microcosms. They detected low-frequency signals corresponding to higher organizational levels in the system. They repeated the experiment with and without a substrate in the microcosm. Since the substrate operates to dampen oscillations in the nutrient concentration of the water column, we would expect that the complete, normal system would show higher levels of organization. The data revealed the expected relationship. They detected more low-frequency signals (i.e., more levels of organization) in the complete system and fewer in the microcosm without a substrate.

This study illustrates how hierarchy theory can be used to

generate testable hypotheses. For example, one might experimentally remove a single population from each of two systems. In one case, the population would be a member of a guild and its removal would involve no change in the basic constraints of the system. As a result, the frequency spectrum before and after removal would show the same number of peaks. In the second experiment, the species would be a keystone species. In this case, removal of the population involves a rupturing of constraints and the frequency spectrum should change radically. In particular, removal of the keystone predator would involve a loss of organization and the frequency spectrum should show fewer peaks.

In fact, an experimental approach to hierarchy theory has already been taken. Van Voris et al. (1980) monitored CO_2 production in terrestrial microcosms. The time series corresponding to each microcosm was transformed into a frequency spectrum. The spectra showed a number of low-frequency peaks indicating higher levels of organization. The appearance of the same low frequency peak in replicate microcosms indicated that the peaks corresponded to organizational levels in the system and were not spurious.

Van Voris and his colleagues predicted that the higher the organizational state of the system, the more stable it should be. This translated into the hypothesis that the microcosms with more frequency peaks would be more stable. They then performed an experiment in which all of the microcosms were perturbed by the addition of a toxic heavy metal. They monitored the loss of a nutrient, calcium, as an indicator of how the systems responded.

All of the microcosms responded to the perturbation by nutrient loss. The systems subsequently recovered and the nutrient loss rates returned to preperturbation levels. The total nutrient loss between perturbation and recovery was taken as a metric of the systems' relative stability.

The test of the hypothesis that organizational complexity

was related to stability was the correlation between the number of peaks shown by each microcosm (i.e., its organizational complexity) and the amount of nutrient lost (i.e., its relative stability). There was a significant positive correlation between the two metrics. The experiment demonstrates that hierarchy theory can make predictions about ecosystem dynamics and that these predictions can be tested experimentally.

CONCLUSIONS

In this chapter, we have explored some of the implications of hierarchy theory for the study of the functional aspects of the ecosystem. The concept of incorporation helped explain how ecological systems come to interact with and control environmental perturbations. Hierarchy theory also gave some insight into what happens when an ecological system becomes unstable. Finally, we showed that functional components of ecosystem behavior can be derived from measures of total system dynamics without explicit reference to arbitrary, tangible components.

Chapter 7 attempted to establish that hierarchy theory is a useful tool in the analysis of population–community phenomena. Chapter 8 has attempted a similar task for process–functional phenomena. Chapter 7 did not demonstrate that functional dynamics such as nutrient cycling could be deduced from the properties of individual isolated populations. Chapter 8 did not demonstrate that community dynamics could be deduced from the functional representation of the ecosystem. It is time to turn our attention to how these two views can be synthesized. We hope that within this synthesis will be found the seeds from which a more unified theory of ecology can be developed.

CHAPTER 9

Ecosystems as Dual Hierarchies

To this point we have shown that the natural world can be profitably viewed as a multilayered system in space and time. This hierarchical conceptualization helps place many phenomena in a better perspective and avoids some of the worst ambiguities associated with the ecosystem concept. Furthermore, the concept seems to be applicable to both population–community and process–functional problems.

We will now turn our attention to the way hierarchy theory helps explain the relationship between the two schools of ecology. Population ecology emphasizes individuals organized into populations, guilds, and communities. It deals with tangible entities and their aggregates. Process–functional ecology emphasizes phenomena such as productivity and nutrient cycling. It seeks functional components that may or may not be species populations.

Simply stated, the problem is that the relationship between organisms in a system and the functions of the system is not always clear. Hierarchy theory provides a framework for understanding this relationship, without reducing one area of emphasis to the other. The resolution of the problem is simply that the two fields are looking at the natural world from different observation sets. MacMahon et al. (1978, 1981) and Webster (1979) have pointed out that at least two independent hierarchies are needed for explaining the full range of ecological phenomena. Much of this chapter can be viewed as an elaboration of their insight.

The higher levels of organization that can be detected in a functional observation set do not always correspond in

186

any simple manner to the levels of organization in the population observation set (MacMahon et al. 1978, 1981). Yet there must be some relationship between the two since, after all, they are simply different observation sets on the same underlying system. We hope that our exploration of the relationship can lead to a more unified theory of ecology which can assimilate the best insights from each point of view.

THE RELATIONSHIP BETWEEN POPULATIONS AND FUNCTIONAL COMPONENTS

In seeking a relationship between populations and functional components, let us first dismiss the hypothesis that there is a simple, one-to-one or even many-to-one mapping of one onto the other. In other words, the functional component need not be a single species (Webster 1979) or even an aggregate of species. There are, of course, important exceptions that we will point out. Nevertheless, it is important to establish that, generally, neither the biotic nor the functional components can be reduced to each other in any simple way.

It is certainly possible to argue, by way of example, (1) that different species can perform the same or similar functions, (2) that the same species can perform several different functions at the same time, and (3) that the same species can perform different functions at different times and places.

It is clear from the common observation of functional redundancy that many species can perform essentially the same ecosystem task. This observation was fundamental to Root's (1967) definition of the "guild." Historical evidence shows that replacement or removal of species, such as the loss of American chestnut to blight, has had essentially no effect on the overall ability of the deciduous forest to fix so-

187

lar energy. In lake ecosystems, primary production appears to be determined by temperature, light, and nutrients. The species assemblage of phytoplankton may change drastically from year to year with little change in productivity. A similar generalization can be made for decomposition processes where rates appear to be determined by abiotic factors. Many different species assemblages of arthropods and microbes may perform the function of decomposition in different microhabitats. The phenomenon of convergent evolution also appears to establish our thesis. As we discuss in detail below, quite different species assemblages may evolve to perform similar functions in geographically separated places. Most importantly, many ecosystem functions, such as nutrient cycling, involve components that are abiotic (e.g., humus–clay aggregates) or combine biotic and abiotic entities (e.g., the rhizosphere). Clearly there is no simple mapping of such components onto assemblages of species.

It is also clear that the same species can perform different functions. The clearest example of this is found in organisms, such as holometabolous insects, that play different functional roles at different stages of their life cycle. In fact, the organism may even change trophic levels, being herbivorous as a larva and predatory as an adult. It was this and similar observations that cast doubts on the concept of trophic level as an assemblage of species populations that could be uniquely identified with an ecosystem function (Koslowsky 1968).

It appears, then, that there is no simple way to reduce all ecosystem functions to unique sets of species that perform the functions. The lack of a simple mapping helps explain why the areal extent of a community may not be immediately relevant to explaining ecosystem functions such as primary production or decomposition (Schultz 1967; MacMahon et al. 1978). A species–area curve might indicate that a large area must be censused to characterize the species composition of a community. Likewise, a gradient analysis might indicate that there are no sharp boundaries between species assemblages across a large area. Yet an ecosystem function such as nutrient cycling can be shown to be operating over a much smaller areal extent within the community. What is more, the rate at which the function changes across the larger area may be more closely correlated with environmental factors such as temperature than it is with the change of species composition. The functional components of the ecosystem exist throughout the gradient even though the species change. In many, perhaps all cases, the community defined by species composition is much larger in spatial extent than ecosystems defined by functions.

That there is no simple relationship between species and ecosystem function helps resolve some apparent dilemmas. Simply because community composition changes over an area does not prevent us from dealing with ecosystem function. Similarly, the definition of an ecosystem at a particular space and time scale does not require that a unique collection of species be identifiable at this scale (Whittaker and Woodwell 1972). It is quite feasible and even reasonable to maintain an individualistic (i.e., Gleasonian) concept of the community and a holistic concept of ecosystem function. Such considerations must make us forever suspicious of attempts to define ecosystem function in terms of a list of species.

The Ecosystem as a Dual Hierarchy

Population–community and process–functional approaches seem to be disjunct and highly dependent on the observation set used. For the population ecologist, the observation set permits one to see individual organisms. The primary entities in the system are biotic and tangible. If one had a long time series of observations, one could detect low-frequency signals that correspond to populations, guilds, communities, and so on. These represent the higher organizational levels within this observation space. The constraints that operate in this observation set are evolutionary principles. The entities (i.e., the organisms) reproduce. The higher levels (i.e., populations) compete and evolve. The relevant constraints result from interactions among the organisms and involve the concept of natural selection.

For the functional ecologist, the observation set can be quite different. Here, the observations are not on the numbers of organisms (tangible and structural) but on the rates at which some process occurs (functional). Indeed, the organism may not even be detectable in the observation set. In the functional observation set, constraints may have little to do with natural selection or competition. Here the constraints are based on conservation of mass and energy.

In the biotic observation set, the tree is a holon. In the functional observation set, tree leaves appear as a high-frequency entity that functions to fix carbon, while tree boles appear as a separate functional entity, combined with detritus, that function to retain and slowly cycle nutrients. In most biotic observation sets, if a perturbation causes the loss of a species, the system has responded unstably. In the functional observation set, the perturbation can cause a stable response since some other population can expand to take over the functional role of the lost population. It is little wonder that confusion results when one ecologist claims that the system is unstable and another claims that it is stable.

Although it is clearly fanciful, a mental analog may prove useful as a heuristic tool. Let us imagine the natural world as a complex solid object, hidden from view behind a pair of opaque screens. The screens are joined along one edge and set at approximately right angles to each other. Let us also imagine that there are lights behind the solid object which cast shadows and images onto the opaque screens. When we stand in front of the screen labeled "population–community," we see projections of the biotic entities in the system. Our observations are on the entities and the questions which arise concern the relationship between the biotic entities and the hierarchical rules that structure assemblages.

If we look at the screen labeled "process–functional," we see a different projection. We see motions and changes which we characterize as processes. The relevant questions involve the rates at which the processes occur and the functional components that perform the functions. In any particular circumstance, what is seen on one screen may be difficult or impossible to explain in terms of what one sees on the other screen, and yet the same object underlies both observation sets (MacMahon et al. 1978, 1981). The relevance of the allegory of the elephant and the blind men continues to force itself upon us.

Complexity and Stability

Failure to distinguish between biotic and functional viewpoints has caused many problems in ecology. Perhaps the most celebrated issue is the relationship between complexity and stability. MacArthur (1955) originally posed the hypothesis in terms of a functional property of ecosystems (i.e., the complexity of energy or material flows). Because of ambiguities in the ecosystem concept (Chapter 1) and the notion that ecosystem dynamics are explicable in terms of populations (Chapter 4), later investigators focused almost entirely on species diversity, and stability defined as the persistence of each species in the system. The result was an attempt to transfer a functional property to the population–community paradigm. Even if species diversity measures an ecologically meaningful property, which it may not (Pielou

191

1977), transferring the concept from the ecosystem to the population is unjustifed. Certainly, tests of the hypothesis have indicated that the approach was unwarranted.

Even if one could argue from a specific observation set that populations were the relevant components of the ecosystem, there are still problems with relating total system stability to diversity at the species level (Woodwell and Smith 1969; McNaughton 1976, 1977). Equivalent species impart a degree of functional redundancy to the ecosystem. A degree of independence from the fine detail of species organization results. Therefore, we can argue for a correlation between ecosystem stability and the functional redundancy of the system. This is quite different from hypothesizing a relationship to species diversity.

The intuitive appeal of the hypothesis is that stability and complexity should be related. Complex ecosystems, such as tropical rain forests and coral reefs, seem to be buffered against exogenous disturbances much better than, for example, tundra or boreal forest ecosystems. However, the generalization has not stood up well to testing. Zaret (1982) found a greater diversity of fish populations in a stable lake system in Panama than in an adjacent but fluctuating riverine system. This observation seemed to confirm that stable environmental conditions are associated with greater diversity. However, when predator fish were introduced into both systems, thirteen of seventeen native fish species were eliminated in the lake and no extinctions were found in the river. Thus, the lake system with greater diversity was not more stable under the perturbation.

It appears that much of the confusion over the stability–diversity issue can be traced to conceptual ambiguities. Not differentiating between population–community and process–functional points of view led to a hypothesis about functional complexity being tested in the population paradigm. A simplistic concept of hierarchy led to the conclusion that complexity should be measured in terms of populations that were assumed to constitute the next lower level of organization.

Alternative State-Space Representations

The observation that there can be alternative representations of the system and that these representations cannot be simply reduced to each other has been placed in more precise terms by Rosen (1977b). His theory calls attention to the way one observes the system and how choices made in setting up an observation set can affect what can be said about the system.

Rosen (1977b) defines a state variable as an observable on the system. It is a mapping from states of the system to real numbers. Thus, if we have a set of states, S, we define the real value function $f:S \rightarrow R$ where R is the set of real numbers and f then represents an observation on the system. In other words, we define the system at a point in time as being in a particular state, which we represent by a set of numbers, such as the number of individuals in each population or the quantity of nutrient in a series of compartments. If the system is in state S, then we characterize that state by a set of state variables:

$$f(s) = f(x_1(s), \ldots, x_n(s)), \tag{9.1}$$

So, f of s is determined by the values of x_i of s. If we now wish to represent the dynamics of the system, we are led to the usual state representation:

$$dx_i/dt = f_i(x_1, \ldots, x_n), \qquad i = 1, \ldots, n. \tag{9.2}$$

What is represented in Equation 9.2 are changes in the state of the system. To define changes, we must establish an equivalence relation, R_f, such that

$$s_1 R_f s_2 \quad \text{iff} \quad f(s_1) = f(s_2).$$

193

This relationship indicates that the two states, s_1 and s_2, are equivalent (i.e., cannot be distinguished) if and only if the state variables have not changed. Conversely, we can detect a change only if $f(s_1)$ does not equal $f(s_2)$ and in turn this can only be true if $x_1(s_1)$ does not equal $x_1(s_2)$ for at least one i. In other words, we can detect a change of state in the system only if one of the state variables we have chosen to describe the system has changed. Our observation set, defined by $f(s)$, conveys limited information simply because it cannot distinguish between states in the same equivalence class, R_f. If the system moves from state s_1 to state s_2 without any change in our state variables, then by definition (Eqs. 9.1 and 9.2) we do not detect any change. Whether or not we detect a change in the system is dependent on the choices we made in setting up our observations.

An example will help at this point. Let us assume that we are interested in community dynamics. We choose as our state variables the number of organisms in each population. If we observe the system at two points in time and all the populations have the same number of individuals, we say that the system has not changed. Clearly, we could set up a second observation set in which we were interested in the age structure of the populations. The state variables would be the number of organisms in each age category. If we observe the system at the same two points in time, we would now conclude that the system had changed: the organisms would be in the next older age category. The important point is that the way we set up our observation set, the way we choose the state variables, determines what will be interesting changes in the system. Unless we record every detail of every organism, an impossible task, then we will ignore some kinds of change in order to focus on others. Depending on this choice, the same system admits of many different representations.

194

RELATIONSHIPS BETWEEN THE HIERARCHIES

Rosen's analysis also indicates that the functional and structural approaches are not orthogonal. In many cases, ecosystem studies measure the rate at which organisms perform a function. On the other hand, natural selection can operate by giving the advantage to organisms that process energy or nutrients more efficiently. The ecosystem shows more than one dimension, and forgetting the difference between the structural and functional dimensions is the source of much confusion. We propose, then, that examining the ecosystem as explicitly composed of more than one dimension is the proper route for defining the relationship between population and ecosystem ecology.

Let us see what is required to compare one observation set, $f(s)$, with a different observation set, $g(s)$ on the same system. In this case, the $g(s)$ is based on a different set of state variables. To accomplish the comparison, we need to define the concept of "closeness." We will say that two states, s_1 and s_2, are "close" if $f(s_1)$ and $f(s_2)$ are close. More technically, we will say that s_2 is f-close to s_1 if

$$[f(s_1) - f(s_2)] < \epsilon. \qquad (9.3)$$

Then, s_1 is a stable point for g if every state f-close to s_1 is also g-close to s_1. In other words, a transition from state s_1 to s_2 cannot be distinguished in either f or g. If s_1 is not a stable point for g, then the neighborhood of s_1 contains an s_2 that is f-close to s_1 but not g-close. In this case, no change is detected in $f(S)$ but a change is detected in $g(S)$. Then, s_1 is a bifurcation point of g with respect to f.

In some observation sets, one can focus entirely on the structural dimension and ignore the functional (e.g., Rosen's $f(s)$). Consider, for example, an observation set on a single population in the laboratory. Here, the experimenter supplies adequate food, nutrients, and so on and controls temperature and other environmental perturbations. Thus, he has removed the functional (mass and energy) constraints from the system. Now the observation set contains only the structural constraints. Competition for space is a phenomenon that is readily seen in this observation set.

Yet some ecologists (e.g., Drury and Nisbit 1973) maintain that competition never occurs. Others argue that competition may occur but that it is very difficult to demonstrate in the field. Resolution of the problem is simple in the context of hierarchy theory. In the multidimensional system, the population is prevented, by mass and energy constraints, from ever reaching population sizes at which competition for space comes into play. The population *can* show competition (constraints on the structural plane) but might *not* show competition under the constraints imposed by the functional plane.

Similar results are often found when laboratory or theoretical results are compared to the field. Thus, Oster (1974) demonstrated the possibility of chaotic behavior in simple population models. The argument was strengthened by comparisons of the model with data from laboratory experiments. Yet O'Neill et al. (1982) argued that such chaotic behavior probably seldom occurs in the context of the normal stochastic environment. In general, many phenomena that occur in models for the number of individuals in a population would never occur if changes in the biomass of the individuals were included in the models. For example, number models seldom allow for functional constraints on input to the system and allow the producer populations to be limited only by their own numbers.

In some observation sets, we can isolate the functional dimension and ignore the structural (MacMahon et al. 1978). For example, consider the measurement of net primary production in a lake. Here, the rate of production is a function of mass-balance constraints. A number of authors have pointed out the close relationship between phosphorus input and productivity (Dillon and Rigler 1974; Vollenweider 1975, 1976; D. W. Schindler 1977, 1978). Others have confirmed the relationship and shown other functional constraints to be operating such as the nitrogen/phosphorus ratio (V. H. Smith 1982) and lake depth (Sakamoto 1866). Individuals come and go, populations come and go, and even guilds of green and blue-green algae come and go, and yet primary production remains relatively constant. One can determine the rate of production without careful consideration of which populations are performing the task.

Just as it is possible to study structural constraints and organization in the laboratory and ignore the functional dimension, it is also possible, in some observation sets, to study the functional dimension and ignore the structural. But just as the problem with the concept of competition shows that focusing on the structural and ignoring the functional can lead to problems, the opposite case can also be made. Consider the task of operating a sewage treatment plant. As long as the sludge contains a full complement of microorganisms, the plant operates as though structural (organism) constraints did not exist. One ignores the organisms and controls temperature and flow rates. But if toxic metals enter the system, the organisms die. Now one can manipulate the temperature controls to no avail. Suddenly the system is not operating in response to functional constraints, but according to biotic constraints. The solution is to remove the metal and reinfuse the sludge with the original complement of organisms.

Although we can imagine observation sets in which one or the other dimension can be isolated, for many problems, constraints arising from both dimensions are likely to play a role. An excellent example is provided by the study of Bot-

197

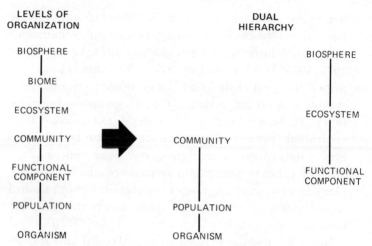

Fig. 9.1. The dual hierarchical structure of ecological systems. To the left is shown the typical levels of organization discussed in Chapter 4. To the right the population–community and process–functional approaches are considered as separate hierarchies.

kin and Levitan (in press) on the Isle Royale ecosystem. Here, the critical factors involve both biotic constraints (wolves interacting with moose) and functional constraints (the rate at which sodium is recycled). It was only when both dimensions of the system were considered that the dynamics could be explained. A similar case arises in deriving predictive equations for terrestrial primary production (Webb et al. 1983). Abiotic constraints will account for much of the variance in the data, but the predictions are greatly improved by including information on the vegetation type. Neither the structural–biotic nor the functional constraints should be ignored in many ecological problems.

Consideration of the ecosystem as a dual hierarchy (MacMahon et al. 1978) also helps clarify some of the confusion that results from adopting the concept of levels of organization. In Figure 9.1, the typical hierarchy is shown on the left. To the right of the figure, the dual hierarchies are separated and can be seen in our present context to be relatively independent of each other.

Alternating Constraints

In the context of the dual hierarchy, it is tempting to speculate on how the two constraint systems interdigitate. Within any given observation set, the system may be structured either by biotic constraints or by mass and energy balance. If something happens to disrupt the existing constraint, can we predict at least what type of constraint would come into play to structure the system?

Based on an anecdotal analysis of examples, we may speculate that the constraint systems alternate. That is, when the system is released from a constraint on one hierarchy, it will next meet a constraint on the other hierarchy. Consider the diatom example presented in Chapter 8. When the diatoms evolved a siliceous shell, they freed themselves from a biotic constraint, predation. They expanded until they were constrained by a mass-balance limitation, inability to extract more silica from the sea water. We might also consider an experiment in which we supplied food to a system which was previously constrained by this resource base. We might then expect the populations to expand until they become constrained by biotic competition.

This speculation suggests what will happen to a system that goes unstable by having its constraint system disrupted. If the constraint system on the other dimension is left intact, the system may develop like a positive feedback system until it is constrained by the next higher (i.e., slower) level on the intact hierarchy.

Dynamic Interfaces between Biotic and Functional Constraints

An ecological system at any level of resolution may be constrained by biotic mechanisms such as competition, or by functional constraints such as lack of water or nutrients. In fact, the system may be limited by both types of constraints simultaneously. Therefore, it is necessary for us to consider how to differentiate between these constraints and how to determine which set of constraints must be considered relative to a specific problem.

To accomplish this, let us assume that the relevant interactions in a system are described by a set of equations for n populations with densities N_i:

$$dN_i/dt = f_i(N_i, N_j, P), \qquad i, j = 1, 2, \ldots, n \qquad (9.4)$$

where P is an array of parameters such as carrying capacities and interaction coefficients. At the time scale at which these Ps remain constant, Equations 9.4 describe the relevant dynamics of the populations. However, the parameters are not really constant. Therefore, we will consider them to be functions of a set of functional components, Y, of the system: $P = h(Y)$. The dynamics of the functional components can then be described by:

$$dY_k/dt = g_k(Y_k, Y_l, R), \qquad k, l = 1, \ldots, m \qquad (9.5)$$

where R is an array of parameters such as rates of nutrient remineralization. Of course, in the present context, we must also allow the possibility that R can vary over time as a function of the biotic components, N: $R = p(N)$.

The equation systems represented by Equations 9.4 and 9.5 can now be considered as alternative representations of the system on the biotic and functional planes. Depending on the nature of the problem, it is conceivable that dP/dt and dR/dt approach zero. In this case both equations are valid representations of different aspects of the system. On the other hand, both P and R may vary over the time scale of the problem and neither approach will work in isolation. Most commonly, we would expect that either P or R could be considered to be constant. If dR/dt is zero, then we can assume that the biotic aspects of the system are relatively constant and the dynamics of the system will follow Equation 9.5. If dP/dt is zero, then we can assume that the functional aspects of the system are relatively constant and the dynamics of interest will be represented by Equation 9.4. These common cases represent problem areas that can legitimately be handled by population–community approaches or by functional approaches.

The dangerous cases arise when one erroneously assumes that the relevant dynamics of the system can be expressed by one approach or the other irrespective of the scale at which one is approaching the system. It is easy to see from the equations that if relevant dynamics at a given scale are described by Equation 9.4 (i.e., P is constant) then it is likely that at the next larger scale, the relevant dynamics will switch to the functional plane and will be concerned with the time behavior of P, which is determined by Equation 9.5.

Is Equilibrium a Relevant Concept?

Equations 9.4 and 9.5 also help clarify the question of whether or not communities and ecosystems can be profitably considered as equilibrium systems. Connell and Sousa (1983) argue that it is practically impossible to prove that a large-scale system is at equilibrium, the proof requiring far longer time series of data than are available. Likens (1983) presents evidence from the Hubbard Brook study which indicates that at least twenty years of continuous records are needed to determine major trends in watershed biogeochemistry. Likens cites Goldman (1981) as indicating that fifteen years of data were needed to demonstrate a decline in transparency of Lake Tahoe. With such long-term records needed to indicate equilibrium or a unidirectional change, it appears to be impractical to deal with the ecosystem as an equilibrium system.

201

It is certainly clear that ecosystems change as climate and other environmental variables change. Hence, many would argue that the ecosystem is essentially a disequilibrium system and that any attempt to describe it in terms of equilibrium leads one into a cul-de-sac.

The controversy carries over into concepts of stability. Most definitions of stability involve a qualitative statement about whether the system returns to an equilibrium following perturbation. To many ecologists, this is an irrelevant concept because no equilibrium exists to which the system can return. Botkin and Sobel (1975) argue that the most relevant concept of stability is one that pictures the system as persistent, changing through time but remaining within defined bounds.

In fact, this convoluted set of problems can be greatly clarified by considering the ecological system as hierarchically structured. The ecological system can be in equilibrium at a lower level in the dynamic hierarchy while remaining far from equilibrium at a higher level. Let us consider Equation 9.4 to represent dynamics at a particular level in an ecological system. The dynamics are constrained by the parameters P, which are themselves functions of a higher level in the system with dynamics described by Equation 9.5.

For the sake of our discussion, let us assume that the system (Eq. 9.4) is globally stable about a single nontrivial equilibrium point, defined by the n-vector, N^*. A perturbation that takes any combination of populations away from N^* will be followed by an asymptotic return to N^* when the perturbation is removed.

Such a representation can be defended if and only if the scale at which one is examining the system shows that the factors that determine P are, in fact, unchanging over the period of observation. This seems a stringent condition and the experienced ecologist will immediately begin to list exceptions. But the point is that the representation is legitimate within these conditions. At short time scales, the N rapidly come into equilibrium, $N^*(Y)$. However, the equilibrium point is itself a dynamic entity at a much longer time scale. Thus, equilibrium and asymptotic stability can be useful and relevant ecological concepts at one level of the hierarchy (i.e., at one time–space scale), even if the system is seen to be at disequilibrium on another, longer time scale.

A similar point is made by Schumm and Lichty (1965). In discussing the dynamics of geomorphology, they point out that the spatiotemporal scale of a problem will determine which dynamics are to be considered in equilibrium and which are to be considered as changing. Over long geologic time scales, the landscape is continuously changing due to uplift and erosion. On this time scale, the individual river must be considered as far from equilibrium. But if the observation set is much shorter, perhaps in years, then it is most useful to view the drainage system as being in equilibrium. Here attention focuses on the negative feedbacks and self-regulation processes that tend to keep the riverbed in relatively constant configuration. Whether or not it is useful to consider the drainage system as being in equilibrium depends on the spatiotemporal scale being considered.

It appears then that much of the controversy over whether ecological systems are at equilibrium is simply a problem of scale. Those who maintain that equilibrium is relevant to ecology are simply looking at the system at a different time–space scale from those who maintain that asymptotic stability and equilibrium are irrelevant. In fact, viewed in the context of hierarchy theory, the problem is not whether the ecological system is unchanging in any absolute sense, but whether it can be considered to be unchanging at the scale of a particular problem. If this assumption is valid, if measurements on the system change by perhaps less than 1 percent over the period of measurement, then the system can be abstracted *as though* it were at equilibrium. The point is that stability and equilibrium should not be considered to be statements about the nature of the universe. They must be considered as statements about observation sets at particular time–space scales.

Communities Structured by Functional Constraints: Convergent Evolution

If the biotic and functional constraint systems are not, in fact, independent of each other, then we should expect to find cases where the constraints on one plane explicitly structure the system on the other plane. Perhaps the most investigated example is convergent evolution (Orians and Solbrig 1977). The concept, which dates back to Grisebach (1872), is simply that functional constraints, operating through climate, soil, and other abiotic factors create a functional context within which evolution operates. If you find two areas that are very similar in their environmental conditions, you may find that the biotic communities have evolved in remarkably similar ways. For our present purposes, convergent evolution shows the interplay of functional constraints and biotic material showing that the biotic material alone does not uniquely specify the system.

Convergent evolution was intensively investigated during the 1970s by comparing the biota of similar climatic areas in southern California and Chile. The basic hypothesis tested in these studies was originally posed by Mooney et al. (1970): one should find greater similarities between community structure in California and Chile (different taxonomic material, similar climates) than between geographically adjacent areas in California (similar taxonomic material, different climates). Early studies on sclerophyll shrubs (Mooney and Dunn 1970) and ant communities (Hunt 1973) seemed to confirm the hypothesis.

Sage (1973) and Fuentes (1976) tested the hypothesis for

lizard communities. They found the hypothesis to be true for patterns of habitat, food, and time of activity. The utilization patterns were parallel in convergent species. The functional constraints appeared to structure the community into definite patterns. Because adjacent areas in California did not show such parallelism, we must conclude that functional constraints played as great a role as biotic material in determining community structure.

Parsons and Moldenke (1975) compared vegetation structure in the Californian and Chilean sites. They found similar patterns in species richness, growth form, leaf duration, leaf size, and spines. The similarities held for both woody and herbaceous vegetation. Once again, the hypothesis seemed to explain the data since adjacent areas in California did not show such parallels.

Taken as a whole, this series of studies indicates the danger of investigating the biotic material in total isolation from functional constraints. Focusing on the biotic material would lead one to believe that similar taxonomic material would evolve into similar communities. However, examination of adjacent plots, for example, woodlots adjacent to grasslands, indicates that this is not true. The process of natural selection operates within a context, a set of constraints. When these constraints limit the biota, then the biotic system will tend to move up to the boundary conditions set by these constraints. Under these circumstances, we must look for the boundary conditions to shape the community.

Functional Redundancy and Flexible Constraints

Functional constraints (e.g., nutrient limitations) do not ordinarily operate directly on individual populations but on functional holons such as guilds and rhizospheres. As a result, the detailed behavior of the individual population is less constrained. As long as the aggregate behaviors of the

205

holon are preserved, the individual component can adapt somewhat independently. This makes it possible for the entire multilevel system to become more tightly coupled to the environment in which it is enmeshed. Selective pressures operating on individual populations are free to operate without endangering some functional role within the total system. Thus, species replacement or changes in dominance in phytoplankton populations (O'Neill and Giddings 1979) can indicate a change in the system viewed on one scale, and yet little or no change at another organizational level of the system (i.e., total primary production is unchanged).

Pattee (1969, 1973) referred to the principle of optimal constraint: selective loss of lower-level detail at the next higher level serves as a flexible constraint on component behaviors. The constraints are optimal in that they preserve the functional roles of holons while allowing flexibility in component behavior. Weiss (1971) referred to such constraints as macrodeterminacy without stereotypy. In essence, individual components can adopt a number of specialized strategies as long as the holon retains its functional role (Makradakis and Weintraub 1971b). The constraint is that the function not be lost, for then the whole organization is jeopardized. Thus, individual insect species may specialize, as long as the guild retains its function.

Ecosystem Function Structured by Population Constraints

Although, in general, an ecosystem function cannot unequivocally be identified with a species population, there are some important exceptions. We must consider these cases since they have important conceptual and practical implications. In these systems, a single species may be the only biotic entity available to perform a function.

Credit must be given to Paine (1966) for developing the concept of keystone species in his intertidal systems. As we discussed in Chapter 7, the removal of a keystone predator

206

may drastically change the species composition of the system. While it is possible that a change in species composition could occur without any change in ecosystem processes, this seems unlikely. Although specific data are lacking, it seems safe to assume that removal of this one species changes the rates of ecosystem functions, even processes in which that species participates only indirectly, for example, decomposition.

We would expect a similar sensitivity of ecosystem function in systems with impoverished biota. In harsh environments, such as tundras or hot springs (Wiegert 1974, 1975), there may be little or no functional redundancy and removal of a species means removal of a function and potential collapse of ecosystem processes. There are also unique habitats where a small group of species dominates the flora, such as mangroves on the coasts of Caribbean islands. We might also expect a similar relationship in unique systems such as the caves described by Barr (1967), Poulson and White (1969), and Culver (1982), which contain only a few species.

Probably the most common case in which ecosystem function can be directly related to a species population occurs in systems where a single species dominates a function. This occurs in systems with large, long-lived species such as the elephant and whale. For example, Larson (1940) has argued for the importance of the American bison in maintaining the short grass prairie. In such cases, the long-lived species can have a controlling influence on the spatial and temporal processing of energy and nutrients and should be included as a functional component of the ecosystem.

Such controlling species often exert their influence over ecosystem processes indirectly by changing constraints in the environment. One need only consider caribou migrations in the tundra or salmon migrations to small streams to recognize the way in which a single species can alter available nutrients.   Cahn (1929) described the changes that occur when carp is introduced into small northern lakes. The vegetation disappears, the water becomes turbid and the fish populations totally change. It seems clear that the rates of ecosystem processes change radically under these circumstances. Cloudsley-Thompson (1975) discusses a number of cases in which vertebrates serve as controlling species. Thus, alligators build impoundments that alter local vegetation and form refuges for aquatic flora and fauna. Similarly, beaver constructions can significantly alter nitrogen retention in northern streams (Naiman and Melillo 1984). In a manner similar to the carp, the hippopotamus alters process rates in African rivers both by adding excreta and by stirring up sediments.

Whenever a controlling species can be identified, advantage should be taken of the situation. Because the population is a tangible entity, measurements taken on the population can greatly clarify the explanation of ecosystem processing. As we pointed out in the last chapter, even if measurements can be taken on total ecosystem function, additional measurements of a single identifiable functional component greatly simplify the task of systems identification.

In many systems, an impact on a single species is difficult to interpret. Increased mortality of a single species may be nicely compensated by competitive release of other, functionally equivalent species. Thus, loss of a species does not imply loss of ecosystem function and collapse of the constraint structure. But whenever a keystone or controlling species is identified, impacts on the single species become immediately interpretable in terms of total system function. Since the methodology for measuring species effects is far better developed than the methodology for detecting total ecosystem changes, the identification of a keystone or controlling species is immediately relevant to monitoring the system for impacts.

The Relationship between Population and Ecosystem Ecology

What results from our consideration of the multidimensionality of ecosystems (MacMahon et al. 1978) is that the dilemma of population versus ecosystem ecology results from a false dichotomy. If we look across the range of observation sets that constitute the subject matter of ecology, we see the ecosystem as at least a dual entity. In one dimension it is structured according to constraints involving organism interaction and natural selection. In another dimension, it is structured according to constraints that involve mass balance and thermodynamics. It is only in unusual observation sets that just one of these dimensions can be seen in isolation.

As a result, a definition of the ecosystem should not overemphasize the dichotomy between the organisms and the abiotic components of the system (Webster 1979). We have shown in Chapter 8 that the ecosystem incorporates many abiotic components. These components are now internal to the system, not part of the environment at all. T.F.H. Allen et al. (1984) provide an extensive discussion on how what we regard as system and what we regard as environment must change as we change the scale of our observations.

In a similar manner, the question, "How large is an ecosystem?" is impossible to answer in general. The structural size for the system, say, a forest, will have little to do with the minimal functional area needed to detect functions such as primary production or nutrient cycling. That minimal area may be a hectare or less anywhere in the forest. In fact, the question of size is particularly inappropriate since both structural and functional mechanisms are involved in ecosystem incorporation. Thus, the ecological system incorporates fire by spatial mechanisms: the landscape is very large relative to the size of the individual fire. But the system incorporates erratic oscillations in nutrient supply by a functional mechanism, that is, nutrient retention and recycling. The spatial scales over which the two mechanisms operate are disarticulated. You cannot predict the spatial scale of one from the other.

Thus, hierarchy theory can serve to clear up points of contention that emanate from the emphasis on either structural or functional aspects of the system. By observing the hierarchical organization of the system, we see that the ecosystem is multidimensional. Ecosystem function may profitably be seen as being influenced by separate sets of constraints that can be characterized as biotic and functional. The organizational structure of the ecosystem may thus be characterized in two dimensions and one dimension cannot simply be reduced to the other (MacMahon et al. 1978). Questions of the primacy of population versus ecosystem ecology disappear. Arguments as to whether ecosystem processes can be "reduced" to population dynamics disappear. We propose, therefore, that hierarchy theory will be an extremely useful tool for preventing false problems from hampering the progress of ecology.

SUMMARY

We have attempted a broad summation of the principles of hierarchy theory in the context of ecosystem analysis. By confining the presentation to specific observation sets, we have shown that the functional structure of the ecosystem can be explicitly extracted from the data, rather than determined by arbitrary designation of components. This provides an operational method for the definition of ecosystem structure.

We have seen that the ecosystem is a dual organization,

arising from structural constraints that operate on organisms and functional constraints that operate on processes. Neither of these dimensions reduces to the other in any simple way and we have shown that significant ambiguities can be introduced when either is ignored.

Although we will not develop the idea further in the present book, we should point out that this pair of hierarchies represents only two of the possible viewpoints we might adopt. Thus, there are other observation sets on the natural world, relatively independent and disjunct from the ones we are considering here, which are both feasible and interesting. At the present time, an awakening interest in spatial patterning (Paine and Levin 1981) and landscape ecology (Forman and Godron 1981; Naveh 1982) indicates that at least a third dimension must be investigated. In these studies, the interest is focused on the way ecological systems are arrayed in space. This perspective is essentially absent in the process–functional viewpoint and represents only a small proportion of population–community studies.

In the context of hierarchy theory, ecosystem organization appears as a system of constraints. We have identified incorporation as a mechanism by which ecosystems escape limitations imposed by a fluctuating environment. We have shown that the ecosystem shows instability whenever the constraint system is broken down. Thus, we have pointed a finger in the direction of how to investigate potential instabilities in the ecosystem. Environmental managers must be careful never to introduce a perturbation that will disturb the system's natural constraint system.

Finally, we have tried to demonstrate that hierarchy theory is more than a concept (Strong 1982). As a scientific theory, it is capable of explaining awkward observations such as saltatorial evolution and competition that never actually happens. We have shown that the theory is capable of generating testable hypotheses about the stability of ecosys-

211

tems. We have even found an example in which the hypothesis was tested showing that organizational complexity is related to stability (Van Voris et al. 1980).

In many respects, the most exciting aspect of hierarchy theory is that it represents a step outside prevailing paradigms. It is a new way of looking at ecosystems. We have tried to demonstrate that this new orientation is indeed a fruitful one that we can recommend as a new way of overcoming old constraints to the progress of ecology.

Literature Cited

Adams, C. C. 1913. *Guide to the Study of Animal Ecology*. Macmillan, New York.

Adams, C. C. 1917. The new natural history—ecology. *Am. Mus. J.* 7:491–494.

Adams, S. M., B. L. Kimmel, and G. R. Ploskey. 1983. Sources of organic matter for reservoir fish production: a trophic-dynamic analysis. *Can. J. Fish. and Aqu. Sci.* 40:1480–1495.

Ahlgren, I. F., and Ahlgren, C. E. 1960. Ecological effects of forest fires. *Bot. Rev.* 26:483–533.

Allee, W. C., and T. Park. 1939. Concerning ecological principles. *Science* 89:166–169.

Allee, W. C., A. E. Emerson, O. Park, T. Park, and K. P. Schmidt. 1949. *Principles of Animal Ecology*. W. B. Saunders, Philadelphia.

Allen, J. A. 1871. Mammals and winter birds of east Florida and a sketch of the bird faunae of Eastern North America. *Bull. Mus. Comp. Zool.* 2:161–450.

Allen, P. M. 1976. Evolution, population dynamics, and stability. *Proc. Nat. Acad. Sci.* 73(3):665–668.

Allen, T.F.H., S. M. Bartell, and J. F. Koonce. 1977. Multiple stable configuratitons in ordination of phytoplankton community change rates. *Ecol.* 58:1076–1084.

Allen, T.F.H., and H. H. Iltis. 1980. Overconnected collapse to higher levels: urban and agricultural origins, a case study. In *Systems Science and Science, Proceedings of the 24th Annual North American Meeting of the Society for*

General Systems Research, pp. 96–103. Society for General Systems Research, San Francisco.

Allen, T.F.H., and J. F. Koonce. 1973. Multivariate approaches to algal stratagems and tactics in systems analysis of phytoplankton. *Ecol.* 54:1234–1246.

Allen, T.F.H., R. V. O'Neill, and T. W. Hoekstra. 1984. Interlevel relations in ecological research and management: some working principles from hierarchy theory. General Technical Report RM-110, United States Department of Agriculture, Rocky Mountain Forest and Range Experiment Station, Fort Collins, Colorado.

Allen, T.F.H., and T. B. Starr. 1982. *Hierarchy: Perspectives for Ecological Complexity*. Univ. of Chicago Press, Chicago.

Ashby, W. R. 1954. *Design for a Brain*. John Wiley and Sons, New York.

Austin, M. P., and B. G. Cook. 1974. Ecosystem stability: a result from an abstract simulation. *J. Theor. Biol.* 45:435–458.

Awramik, S. M. 1971. Precambrian columnar stromatolite diversity: reflection of metazoan appearance. *Science* 174:825–827.

Awramik, S. M. 1981. The pre-phanerozoic biosphere—three billion years of crises and opportunities. In *Biotic Crisis in Ecological and Evolutionary Time*, ed. M. H. Nitecki, pp. 83–102. Academic Press, New York.

Barr, T. C. 1967. Observations on the ecology of caves. *Amer. Nat.* 101:475–491.

Bertalanffy, L. von. 1975. *Perspectives on General Systems Theory*. Braziller, New York.

Boling, R. H., R. C. Petersen, and K. W. Cummins. 1975. Ecosystem model for small woodland streams. In *Systems Analysis and Simulation in Ecology*, ed. B. C. Patten, Volume 3, pp. 183–204. Academic Press, New York.

Bormann, F. H., and G. E. Likens. 1967. Nutrient cycling. *Science* 155:424–429.

Bormann, F. H., and G. E. Likens. 1979a. *Pattern and Process in a Forested Ecosystem*. Springer-Verlag, New York.

Bormann, F. H., and G. E. Likens. 1979b. Catastrophic disturbance and the steady state in northern hardwood forests. *Amer. Sci.* 67:660–669.

Bosserman, R. W. 1979. The hierarchical integrity of *Utricula*-Periphyton microecosystems. Ph.D. diss., Univ. of Georgia, Athens.

Bossert, A. K., M. A. Jasieniuk, and E. A. Johnson. 1977. Levels of organization. *Bioscience* 27:82.

Botkin, D. B., ed. *Toward a Science of the Biosphere*. Committee on Planetary Biology and Chemical Evolution, National Academy of Sciences, Washington, D.C. (In press.)

Botkin, D. B., and R. E. Levitan. Wolves, moose, and trees: an age-specific trophic-level model of Isle Royale National Park. *Ecology*, in press.

Botkin, D. B., and M. J. Sobel. 1975. Stability in time-varying ecosystems. *Amer. Nat.* 109:625–646.

Braun, E. L. 1950. *Deciduous Forests of Eastern North America*. Blakiston, Philadelphia.

Braun-Blanquet, J. 1932. *Plant Sociology: The Study of Plant Communities*. Translated by G. D. Fuller and H. S. Coward. McGraw-Hill, New York.

Bray, J. R. 1958. Notes toward an ecologic theory. *Ecol.* 39:770–776.

Bray, J. R., and J. T. Curtis. 1957. An ordination of the upland forest communities of southern Wisconsin. *Ecol. Monogr.* 27:325–349.

Bronowski, J. 1973. *The Ascent of Man*. Little, Brown, and Co., Boston.

Bronowski, J. 1977. *A Sense of the Future*. MIT Press, Cambridge.

Bunge, M. 1959a. Do the levels of science reflect the levels of being? In *Metascientific Queries*, chap. 5. C. C. Thomas, Springfield, Ill.

Bunge, M. 1959b. *Causality*. Harvard Univ. Press, Cambridge.

Bunge, M. 1969. The metaphysics, epistemology, and methodology of levels. In *Hierarchical Structures*, ed. L. L. Whyte, A. G. Wilson, and D. Wilson, pp. 17–26. Elsevier, New York.

Cahn, A. R. 1929. The effect of carp on a small lake: the carp as a dominant. *Ecol.* 10:271–274.

Cain, S. A., and G. M. de Castro. 1959. *Manual of Vegetation Analysis*. Harper and Row, New York.

Cairns-Smith, A. G. 1971. *The Life Puzzle*. Univ. of Toronto Press, Toronto.

Candolle, A.P.A. de. 1874. *Constitution dans le règne végétal de groupes physiologiques applicables à la géographie ancienne et moderne*. Archives des Sciences Physiques et Naturelles, Geneva.

Carney, J. H., D. L. DeAngelis, R. H. Gardner, J. B. Mankin, and W. M. Post. 1981. Calculation of probabilities of transfer, recurrence intervals, and positional indices for linear compartment models. ORNL/TM-7379. Oak Ridge National Laboratory, Oak Ridge.

Carpenter, J. R. 1939. The biome. *Amer. Midl. Nat.* 21:75–91.

Child, G. I., and H. H. Shugart. 1972. Frequency response analysis of magnesium cycling in a tropical forest system. In *Systems Analysis and Simulation in Ecology*, ed. B. C. Patten, Volume 1, pp. 103–134. Academic Press, New York.

Clarke, B. C. 1975. The causes of biological diversity. *Sci. Amer.* 233(2):50–60.

Clarke, B. C. 1976. The ecological geneticis of host–parasite relationships. In *Genetic Aspects of Host-Parasite Relation-*

ships, ed. A.E.R. Taylor and R. Muller, pp. 87–103. Blackwell, Oxford.

Clements, F. E. 1916. *Plant Succession: An Analysis of the Development of Vegetation.* Carnegie Inst. Wash. Publ. 242.

Clements, F. E. 1936. Nature and structure of the climax. *J. Ecol.* 24:252–284.

Cloudsley-Thompson, J. 1975. *Terrestrial Environments.* John Wiley and Sons, New York.

Cole, L. C. 1957. Sketches of general and comparative demography. *Cold Spring Harbor Symp. Quant. Biol.* 22:1–15.

Colinvaux, P. A. 1973. *Introduction to Ecology.* John Wiley and Sons, New York.

Collier, B. D., G. W. Cox, A. W. Johnson, and C. P. Miller. 1973. *Dynamic Ecology.* Prentice-Hall, Englewood Cliffs, N.J.

Connell, J. H. 1970. A predator–prey system in the marine intertidal region, I: *Balanus glandula* and several predatory species of *Thais. Ecol. Monogr.* 40:49–78.

Connell, J. H., and E. Orias. 1964. The ecological regulation of species diversity. *Amer. Nat.* 98:387–414.

Connell, J. H., and W. P. Sousa. 1983. On the evidence needed to judge ecological stability or persistence. *Amer. Nat.* 121:789–824.

Conrad, M. 1976. Patterns of biological control in ecosystems. In *Systems Analysis and Simulation in Ecology*, ed. B. C. Patten, Volume IV, pp. 430–456. Academic Press, New York.

Cooper, H.S.F. 1976. *The Search for Life on Mars.* Holt, Rinehart, and Winston, New York.

Cooper, W. S. 1926. The fundamentals of vegetational change. *Ecol.* 7:391–413.

Cowles, H. C. 1899. The ecological relations of the vegetation on the sand dunes of Lake Michigan. *Bot. Gaz.* 27:95–117.

217

LITERATURE CITED

Cowles, H. C. 1901. The physiographic ecology of Chicago. *Bot. Gaz.* 31:73–108.

Cowles, H. C. 1904. The work of the year 1903 in ecology. *Science* 19:879–895.

Culver, D. C. 1982. *Cave Life: Evolution and Ecology.* Harvard Univ. Press, Cambridge.

Cummins, K. W. 1974. Structure and function of stream ecosystems. *Bioscience* 24:631–641.

Cummins, K. W., R. C. Petersen, F. O. Howard, J. C. Wuycheck, and V. I. Holt. 1973. The utilization of leaf litter by stream detritovores. *Ecol.* 54:336–345.

Curtis, J. T. 1959. *The Vegetation of Wisconsin: An Ordination of Plant Communities.* Univ. of Wisconsin Press, Madison.

Curtis, J. T., and R. P. McIntosh. 1951. The interrelations of certain analytic and synthetic characters. *Ecol.* 31:434–455.

Dale, M. B. 1970. Systems analysis and ecology. *Ecol.* 51:2–16.

Dammerman, K. W. 1948. The fauna of Krakatau 1883–1933. *Verh. K. Ned. Akad. Wet. afd. Natuurk* 44:1–594.

Daubenmire, R. 1968. *Plant Communities: A Textbook of Plant Synecology.* Harper and Row, New York.

Davis, M. B. 1976. Pleistocene biogeography of temperate deciduous forests. *Geosci. Man* 13:13–26.

DeAngelis, D. L. 1975a. Stability and connectance in food web models. *Ecol.* 56:238–243.

DeAngelis, D. L. 1975b. Estimation of predator–prey limit cycles. *Bull. Math. Biol.* 37:291–299.

DeAngelis, D. L. 1980. Energy flow, nutrient cycling, and ecosystem resilience. *Ecol.* 61:764–771.

DeAngelis, D. L., W. M. Post, and C. C. Travis. 1986. *Positive Feedback in Natural Systems.* Springer-Verlag, New York.

Denbigh, R. G. 1975. *An Inventive Universe*. Braziller, New York.

Diamond, J. M. 1975. The island dilemma: lessons of modern biogeographic studies for the design of natural reserves. *Biol. Conserv.* 7:129–146.

Dillon, P. J., and F. H. Rigler. 1974. The chlorophyll–phosphorus relationship in lakes. *Limnol. Oceanogr.* 19:767–773.

Dokuchaev, V. V. 1883. *Tchernozeme de la Russie d'Europe.* St. Petersburg.

Drury, W. H., and C. T. Nisbit. 1971. Interrelations between developmental models in geomorphology, plant ecology, and animal ecology. *Gen. Sys.* 16:57–68.

Drury, W. H., and C. T. Nisbet. 1973. Succession. *J. of the Arnold Arboretum* 54:331–368.

Du Rietz, G. E. 1929. The fundamental units of vegetation. *Proc. Int. Congr. of Plant Sci.* 1:623–627.

Dwyer, R. L., and K. T. Perez. 1983. An experimental examination of ecosystem linearization. *Amer. Nat.* 121:305–323.

Eddington, A. 1939. *The Philosophy of Physical Science*. Cambridge Univ. Press, Cambridge.

Egerton, F. N. 1973. Changing concepts of the balance of nature. *Quart. Rev. Biol.* 48:322–350.

Egerton, F. N. 1976. Ecological studies and observations before 1900. In *Issues and Ideas in America*, ed. B. J. Taylor and T. J. White, pp. 311–351. Univ. of Oklahoma Press, Norman.

Egler, F. E. 1942. Vegetation as an object of study. *Phil. of Sci.* 9:245–260.

Erhlich, P. R., and L. C. Birch. 1967. The "balance of nature" and "population control." *Amer. Nat.* 101:97–107.

Eigen, M., and P. Schuster. 1979. *The Hypercycle*. Springer-Verlag, Berlin.

Eigen, M., W. C. Gardiner, P. Schuster, and R. Winkler-Oswatitsch. 1981. The origin of genetic information. *Sci. Amer.* 244(4):88.

Elton, C. S. 1927. *Animal Ecology.* Macmillan, New York.

Emanuel, W. R., D. C. West, and H. H. Shugart. 1978. Spectral analysis and forest dynamics: long-term effects of environmental perturbations. In *Time Series and Ecological Processes*, ed. H. H. Shugart, pp. 195–210. Society for Industrial and Applied Mathematics, Philadelphia.

Engelberg, J., and L. L. Boyarsky. 1979. The noncybernetic nature of ecosystems. *Amer. Nat.* 114:317–324.

Evans, F. C. 1956. Ecosystem as the basic unit in ecology. *Science* 123:1127–1128.

Feeney, P. 1975. Biochemical coevolution between plants and their insect herbivores. In *Coevolution of Animals and Plants*, ed. L. E. Gilbert and P. H. Raven, pp. 3–19. Univ. of Texas Press, Austin.

Feeney, P. 1982. Coevolution of plants and insects. In *Current Themes in Tropical Sciences, 2: Natural Products for Innovative Pest Management*, ed. T. R. Odhiambo, chap. 11. Pergamon Press, Oxford.

Feibleman, J. K. 1954. Theory of integrative levels. *Brit. J. Phil. Sci.* 5:59–66.

Finerty, J. P. 1980. *The Population Ecology of Cycles in Small Mammals.* Yale Univ. Press, New Haven.

Fish, D., and S. R. Carpenter. 1982. Leaf litter and larval mosquito dynamics in tree-hole ecosystems. *Ecol.* 63:283–288.

Fisher, R. A. 1930. *The Genetic Theory of Natural Selection.* Clarendon Press, Oxford.

Forbes, E. 1844. Report on the mollusca and radiata of the Aegean Sea. *Rep. Brit. Assoc. Adv. Sci.* 1844:130–193.

Forbes, S. A. 1887. The lake as a microcosm. *Bull. Sci. Assoc. of Peoria, Illinois* 1887:77–87.

Forcier, L. K. 1975. Reproductive strategies and the co-occurrence of climax tree species. *Science* 189:808–810.

Forman, R.T.T., and M. Godron. 1981. Patches and structural components for a landscape ecology. *Bioscience* 31:733–740.

Fredericks, K. 1958. A definition of ecology and some thoughts about basic concepts. *Ecol.* 39:154–159.

Frey, D. G. 1953. Regional aspects of the late-glacial and post-glacial pollen succession of southeastern North Carolina. *Ecol. Monogr.* 23:289–313.

Fuentes, E. R. 1976. Ecological convergence of lizard communities in Chile and California. *Ecol.* 57:3–17.

Gallup, J. P., and S. V. Benson. 1979. Phase locking in the oscillations leading to turbulence. In *Pattern Formation by Dynamic Systems and Pattern Recognition*, ed. H. Haken, pp. 74–80. Springer-Verlag, New York.

Gardner, M. R., and W. R. Ashby. 1970. Connectance of large dyamical (cybernetic) systems: critical values for stability. *Nature* 228:784.

Gause, G. F. 1934. *The Struggle for Existence*. Williams and Wilkins, Baltimore.

Geisler, S. 1926. Soil reactions in relation to plant successions in the Cincinnati region. *Ecol.* 7:163–184.

Gerard, R. W. 1969. Hierarchy, entiation, and levels. In *Hierarchical Structure*, ed. L. L. Whyte, A. G. Wilson, and D. Wilson, pp. 215–228. Elsevier, New York.

Gill, A. M. 1975. Fire and the Australian flora: a review. *Austr. For.* 38:3–25.

Gingerich, P. D. 1983. Rates of evolution: effects of time and temporal scaling. *Science* 222:159–161.

Gleason, H. A. 1917. The structure and development of the plant association. *Bull. Torrey Bot. Club* 43:463–481.

Gleason, H. A. 1926. The individualist concept of the plant association. *Bull. Torrey Bot. Club* 53:7–26.

Gleason, H. A. 1936. Is the synusia an association? *Ecol.* 17:444–451.

Gleason, H. A. 1939. The individualistic concept of the plant association. *Amer. Midl. Nat.* 21:92–110.

Goh, B.-S. 1980. *Management and Analysis of Biological Populations.* Elsevier, New York.

Goldman, C. R. 1981. Lake Tahoe: two decades of change in a nitrogen deficient oligotrophic lake. *Verh. Int. Verein. Limnol.* 21:45–70.

Goodall, D. W. 1954. Objective methods for the classification of vegetation. *Austr. J. Bot.* 2:304–324.

Goodall, D. W. 1974. Problems of scale and detail in ecological modeling. *J. Env. Manag.* 2:149–157.

Goodwin, B. C. 1963. *Temporal Organization in Cells.* Academic Press, London.

Gould, S. J. 1980. Is a new and general theory of evolution emerging? *Paleobiology* 6:119–130.

Graham, S. A. 1925. The felled tree trunk as an ecological unit. *Ecol.* 6:397–411.

Grene, M. 1967. Biology and the problem of levels of reality. *New Scholasticism* 41:427–449.

Grene, M. 1969. Hierarchy: one word, how many concepts? In *Hierarchical Structures,* ed. L. L. Whyte, A. G. Wilson, and D. Wilson, pp. 56–58. Elsevier, New York.

Grisebach, A. 1872. *Die Vegetation der Erde nach ihrer klimatischen Anordnung.* Englemann, Leipzig.

Gutman, H. 1969. Structure and function in living systems. In *Hierarchical Structures,* ed. L. L. Whyte, A. G. Wilson, and D. Wilson, pp. 229–230. Elsevier, New York.

Guttman, B. S. 1976. Is "levels of organization" a useful biological concept? *Bioscience* 26:112–113.

Hairston, N. G., F. E. Smith, and L. B. Slobodkin. 1960. Communitiy structure, population control, and competition. *Amer. Nat.* 94:421–425.

Hall, A. D., and R. E. Fagan. 1956. Definition of a system. *Gen. Sys.* 1:18–29.

Hamilton, W. D. 1980. Sex versus non-sex versus parasite.

Oikos 35:282–290. Reprinted in *The Mathematical Theory of the Dynamics of Biological Populations II*, ed. R. W. Hiorns and D. Cooke, pp. 139–155. Academic Press, London.

Hamilton, W. D. 1983. Pathogens as causes of genetic diversity in their host populations. In *Population Biology of Infectious Diseases*, ed. R. M. Anderson and R. M. May, pp. 269–296. Springer-Verlag, Heidelberg.

Hammons, A., ed., 1981. Methods for ecological toxicology: a critical review of laboratory multispecies tests. ORNL-5708, EPA-560/11-80-026. Oak Ridge National Laboratory, Oak Ridge.

Haskell, E. F. 1940. Mathematical systematization of environment, organism, and habitat. *Ecol.* 21:1–16.

Heatwole, H. 1971. Marine-dependent terrestrial biotic communities on some cays in the Coral Sea. *Ecol.* 52:363–366.

Heatwole, H., and R. Levins. 1972. Trophic structure, stability, and faunal change during recolonization. *Ecol.* 53:531–534.

Heatwole, H., and R. Levins. 1973. Biogeography of the Puerto Rican Bank: species-turnover on a small cay, Cayo Ahogado. *Ecol.* 54:1042–1055.

Hill, J., and S. L. Durham, 1978. Input, signals, and controls in ecosystems. In *Proc. Conf. on Acoustics, Speech, and Signal Processing*, pp. 391–397. Institute of Electrical and Electronic Engineeers, New York.

Hill, J., and R. G. Wiegert. 1980. Microcosms in ecological research. In *Microcosms in Ecological Research*, ed. J. P. Giesey, pp. 138–163. Savannah River Ecology Laboratory, Aiken.

Hoyle, F. 1977. *Ten Faces of the Universe*. W. H. Freeman, San Francisco.

Humboldt, A. von. 1807. *Ideenzu einer Geographie der Pflan-*

gen nebat einem Naturgemalde der Tropenlander. Tubingen.

Hunt, J. H. 1973. Comparative ecology of ant communities in Mediterranean regions of California and Chile. Ph.D. diss., Univ. of California, Berkeley.

Hutchinson, G. E. 1948. Circular causal systems in ecology. *Ann. N.Y. Acad. Sci.* 50:221–246.

Hutchinson, G. E. 1953. The concept of pattern in ecology. *Proc. Nat. Acad. Sci. Philadelphia* 105:1–12.

Hutchinson, G. E. 1978. *An Introduction to Population Ecology.* Yale Univ. Press, New Haven.

Huxley, J. 1943. *Evolution, the Modern Synthesis.* Harper and Bros., New York.

Iglich, E. 1975. Age structure of red, black, and scarlet oaks, sourwood, sourgum, and tulip tree populations on Watershed 18 at Coweeta. M.S. thesis, Univ. of Georgia, Athens.

Innis, G. S. 1978. *Grassland Simulation Model.* Springer-Verlag, New York.

Jacob, F. 1976. *The Logic of Life: A History of Heredity.* Random House, New York.

Jacot, A. P. 1936. Soil structure and soil biology. *Ecol.* 17:359–379.

Jaenike, J. 1978. An hypothesis to account for the maintenance of sex within populations. *Evolutionary Theory* 3:191–194.

Jansen, D. 1980. When is it coevolution? *Evol.* 34:611–612.

Jantsch, E. 1980. *The Self-Organizing Universe.* Pergamon Press, New York.

Jenny, H. 1930. *A Study of the Influences of Climate upon the Nitrogen and Organic Matter Content of the Soil.* Missouri Agricultural Exp. Station Bulletin No. 152.

Johnson, L. 1981. The thermodynamic origin of ecosystems. *Can. J. Fish. and Aqu. Sci.* 38:571–590.

Kerner, E. H. 1957. A statistical mechanics of interacting biological species. *Bull. Math. Biophysics* 19:121–146.

Kerner, E. H. 1959. Further considerations on the statistical mechanics of biological associations. *Bull. Math. Biophysics* 21:217–255.

Klir, G. J. 1969. *An Approach to General Systems Theory.* Van Nostrand Reinhold, New York.

Knoll, A. H., R. V. O'Neill, D. L. DeAngelis, and L. B. Slobodkin. Interfacing the ecological and evolutionary time scales. Typescript.

Koestler, A. 1967. *The Ghost in the Machine.* Macmillan, New York.

Koestler, A. 1969. Beyond atomism and holism—the concept of the holon. In *Beyond Reductionism*, ed. A. Koestler and J. R. Smythies, pp. 192–232. Hutchinson, London.

Koppen, W. 1884. Die Warmezonen der Erde, nach der Dauer der Heissen, Gemassigten, und Kalten Zeit, und nach der Wirkung der Warme auf die Organisch Welt betrachtet. *Meteorologische Zeitschrift* 1:215–226.

Koslovsky, D. G. 1968. A critical evaluation of the trophic level concept. *Ecol.* 49:48–60.

Krajina, V. 1960. Ecosystem classification of forests. *Silva Fennica* 105:107–110.

Lane, P. A., G. H. Lauff, and R. Levins. 1975. The feasibility of using a holistic approach in ecosystem analysis. In *Ecosystem Analysis and Prediction*, ed. S. A. Levin, pp. 111–130. Society for Industrial and Applied Mathematics, Philadelphia.

Larsen, J. A. 1922. Effect of removal of the virgin white pine stand upon the physical factors of site. *Ecol.* 3:302–305.

Larson, F. 1940. The role of the bison in maintaining the short grass plains. *Ecol.* 21:113–121.

Lassen, N. A., and O. Henriksen. 1983. Tracer studies of

peripheral circulation. In *Tracer Kinetics and Physiologic Modelling*, ed. R. M. Lambrecht and A. Rescigno, pp. 235–297. Springer-Verlag, New York.

Leake, C. D. 1969. Historical aspects of the concept of organizational levels of living material. In *Hierarchical Structures*, ed. L. L. Whyte, A. G. Wilson, and D. Wilson, pp. 147–159. Elsevier, New York.

Levandowsky, M., and B. S. White. 1977. Randomness, times series, and the evolution of biological communities. *Evol. Biol.* 10:69–161.

Levin, S. A. 1975. Editor's comment. In *Niche Theory and Application*, ed. R. H. Whittaker and S. A. Levin, pp. 70–72. Hutchinson Ross, Stroudsburg, Pa.

Levin, S. A. 1983. Some approaches to the modelling of coevolutionary interactions. In *Coevolution*, ed. M. H. Nitecki, pp. 21–65. Univ. of Chicago Press, Chicago.

Levin, S. A., ed. 1976. *Ecological Theory and Ecosystem Models.* The Institute of Ecology, Indianapolis.

Levins, R. 1970. Complex systems. In *Toward a Theoretical Biology*, ed. C. H. Waddington, Volume 3, pp. 73–88. Aldine-Atherton, Chicago.

Levins, R. 1973. The limits of complexity. In *Hierarchy Theory*, ed. H. H. Pattee, pp. 109–127. Braziller, New York.

Levins, R. 1974. The qualitative analysis of partially specified systems. *Ann. N.Y. Acad. Sci.* 231:123–138.

Libby, W. F. 1955. *Radiocarbon Dating*. Univ. of Chicago Press, Chicago.

Likens, G. E. 1983. A priority for ecological research. *Bull. Ecol. Soc. of Amer.* 64:234–243.

Likens, G. E., F. H. Bormann, N. M. Johnson, D. W. Fisher, and R. S. Pierce. 1970. Effects of forest cutting and herbicide treatment on nutrient budgets in the Hubbard Brook Watershed ecosystem. *Ecol. Monogr.* 40:23–47.

Lindeman, R. L. 1942. The trophic dynamic aspect of ecology. *Ecol.* 23:399–418.

Livingston, D. A. 1955. Some pollen profiles from arctic Alaska. *Ecol.* 36:587–600.

Lotka, A. J. 1925. *Elements of Physical Biology.* Williams and Wilkins, Baltimore.

Loucks, O. L. 1970. Evolution of diversity, efficiency, and community stability. *Amer. Zool.* 10:17–25.

Ludwig, D., D. D. Jones, and C. S. Holling. 1978. Qualitative analysis of insect outbreak systems: the spruce budworm and forest. *J. Anim. Ecol.* 47:315–332.

Lugo, A. E., M. Sell, and S. C. Snedaker. 1976. Mangrove ecosystem analysis. In *Systems Analysis and Simulation in Ecology*, ed. B. C. Patten, Volume IV, pp. 113–145. Academic Press, New York.

MacArthur, R. H. 1955. Fluctuations of animal populations and a measure of community stability. *Ecol.* 36:533–536.

MacArthur, R. H. 1971. Patterns of terrestrial bird communities. In *Avian Biology*, ed. D. S. Farner and J. R. King, pp. 189–221. Academic Press, New York.

MacArthur, R. H. 1972. Strong or weak interactions? *Trans. Conn. Acad. Arts and Sci.* 44:177–188.

McColloch, J. W., and W. P. Hayes, 1922. The reciprocal relation of soil and insects. *Ecol.* 3:288-301.

MacFadyen, A. 1975. Some thoughts on the behavior of ecologists. *J. Anim. Ecol.* 44:351–363.

McIntire, C. D., J. A. Colby, and J. D. Hall. 1975. The dynamics of small lotic ecosystems: a modeling approach. *Verh. Int. Verein. Limnol.* 19:1599–1609.

McIntire, C. D., and J. A. Colby. 1978. A hierarchical model of lotic ecosystems. *Ecol. Monogr.* 48:167–190.

McIntosh, R. P. 1976. Ecology since 1900. In *Issues and Ideas in America,* ed. B. I. Taylor and T. J. White (eds.), pp. 353–372. Univ. of Oklahoma Press, Norman.

McIntosh, R. P. 1980. The background and some problems of theoretical ecology. *Synthèse* 43:195–255.

McIntosh, R. P. 1985. *The Background of Ecology.* Cambridge Univ. Press, Cambridge.

McKenzie, L. 1966. Matrices with dominant diagonals and economic theory. In *Proc. Symp. on Mathematical Methods in Social Sciences*, pp. 47–62. Stanford Univ. Press, Stanford.

MacMahon, J. A., D. L. Phillips, J. V. Robinson, and D. J. Schimpf. 1978. Levels of biological organization: an organism-centered approach. *Bioscience* 28:700–704.

MacMahon, J. A., D. J. Schimpf, D. C. Andessen, K. G. Smith, and R. L. Bayn. 1981. An organism-centered approach to some community and ecosystem concepts. *J. Theor. Biol.* 88:287–307.

McMurtie, R. F. 1975. Determinants of stability of large, randomly connected systems. *J. Theor. Biol.* 50:1–11.

McNaughton, S. J. 1976. Serengeti migratory wildebeest facilitation of energy flow by grazing. *Science* 191:92–94.

McNaughton, S. J. 1977. Diversity and stability of ecological communities: a comment on the role of empiricism in ecology. *Amer. Nat.* 111:515–525.

Maelzer, D. A. 1965. Environment, semantics, and systems theory in ecology. *J. Theor. Biol.* 8:395–402.

Major, J. 1951. A functional, factorial approach to plant ecology. *Ecol.* 32:392–412.

Major, J. 1969. Historical development of the ecosystem concept. In *The Ecosystem Concept in Natural Resource Management*, ed. G. M. Van Dyne, pp. 9–22. Academic Press, New York.

Makridakis, S., and C. Faucheux. 1973. Stability properties of general systems. *Gen. Sys.* 18:3–12.

Makridakis, S., and E. R. Weintraub. 1971a. On the synthesis of general systems, Part I: The probability of stability. *Gen. Sys.* 16:43–50.

Makridakis, S., and E. R. Weintraub. 1971b. On the synthesis of general systems, Part II: Optimal system size. *Gen. Sys.* 16:51–54.

Margalef, R. 1963. On certain unifying principles in ecology. *Amer. Nat.* 97:357–374.

Margalef, R. 1968. *Perspectives in Ecological Theory.* Univ. of Chicago Press, Chicago.

Margulis, L. 1971. Symbiosis and evolution. *Sci. Amer.* 225:48–57.

Margulis, L. 1981. *Symbiosis in Cell Evolution.* W. H. Freeman, San Francisco.

Marples, T. G. 1966. A radionuclide tracer study of arthropod food chains in *Spartina* salt-marsh ecosystem. *Ecol.* 47:270–277.

Maruyama, M. 1963. The second cybernetics: deviation-amplifying mutual causal processes. *Amer. Sci.* 51:164–179.

Mason, H. L., and J. H. Langeheim. 1957. Language and the concept of environment. *Ecol.* 38:325–340.

Matsuno, K. 1978. Evolution of dissipative systems: a theoretical basis for Margalef's principle of ecosystem. *J. Theor. Biol.* 70:23–31.

Mattson, M. J., and N. D. Addy. 1975. Phytophagous insects as regulators of forest primary production. *Science* 190:515–522.

May, R. M. 1972. Will a large complex system be stable? *Nature* 238:413–414.

May, R. M. 1973a. *Stability and Complexity in Model Ecosystems.* Princeton Univ. Press, Princeton.

May, R. M. 1973b. Qualitative stability in model ecosystems. *Ecol.* 54:638–641.

May, R. M. 1974. Ecosystem patterns in randomly fluctuating environments. *Progress in Theor. Biol.* 3:1–50.

May, R. M., and G. F. Oster. 1976. Bifurcations and dynamic complexity in simple models. *Amer. Nat.* 110:573–599.

Maynard Smith, J. 1974. *Models in Ecology.* Cambridge Univ. Press, Cambridge.

Merriam, C. H. 1890. Results of a biological survey of the San Francisco Mountain region and desert of the Little Colorado, Arizona. *North American Fauna* 3:1–136.

Merriam, C. H. 1898. *Life Zones and Crop Zones of the United States.* U.S. Dept. Agr. Bull. 10. Washington, D.C.

Mesarovic, M. D., and D. Macko. 1969. Foundations for a scientific theory of hierarchical systems. In *Hierarchical Structures*, ed. L. L. Whyte, A. G. Wilson, and D. Wilson, pp. 29–50. Elsevier, New York.

Mesarovic, M. D., D. Macko, and Y. Takahara. 1970. *Theory of Hierarchical Multilevel Systems.* Academic Press, New York.

Mobius, K. 1877. Die Auster und die Austernwirtschaft. *Trans. Report U.S. Fish. Comm.* 1880:683–751.

Monod, J. 1972. *Chance and Necessity.* Random House, New York.

Mooney, H. A., and E. L. Dunn. 1970. Convergent evolution of Mediterranean climate evergreen sclerophyll shrubs. *Evol.* 24:292–303.

Mooney, H. A., E. L. Dunn, F. Shropshire, and L. Song. 1970. Vegetation comparisons between the Mediterranean climatic areas of California and Chile. *Flora* 159:480–496.

Mooney, H. A., T. M. Bonnicksen, N. L. Christensen, J. E. Lotan, and W. A. Reiners, eds. 1981. Fire regimes and ecosystem properties. Technical Report wo-26. U.S. Forest Service, Washington, D.C.

Morowitz, H. J. 1968. *Energy Flow in Biology.* Academic Press, New York.

Mount, A. B. 1964. The interdependence of the eucalypts

and forest fires in southern Australia. *Austr. For.* 28:166–172.

Mueller-Dumbois, D., and H. Ellenberg. 1974. *Aims and Methods of Vegetation Ecology.* John Wiley and Sons, New York.

Muller-Herold, U. 1983. What is a hypercycle? *J. Theor. Biol.* 102:569–584.

Murdoch, W. W. 1966. Community structure, population control, and competition—a critique. *Amer. Nat.* 100:219–226.

Murdoch, W. W. 1979. Predation and the dynamics of prey populations. *Fortschn. Zool.* 25:295–310.

Mutch, R. W. 1970. Wildland fires and ecosystems: a hypothesis. *Ecol.* 51:1046–1051.

Naiman, R. J., and J. M. Mellilo. 1984. Nitrogen budget of a subarctic stream altered by beaver (*Castor canadensis*). *Oecologia* 62:150–155.

Naveh, Z. 1982. Landscape ecology as an emerging branch of human ecosystem science. *Adv. Ecol. Res.* 12:189–237.

Newbold, J. D., P. J. Mulholland, J. W. Elwood, and R. V. O'Neill. 1982. Organic carbon spiralling in stream ecosystems. *Oikos* 38:266–272.

Nicholis, G., and I. Prigogine. 1977. *Self-Organization in Non-Equilibrium Systems: From Dissipative Structures to Order through Fluctuations.* Wiley Interscience, New York.

Noble, I. R. and R. O. Slatyer. 1980. The use of vital attributes to predict successional changes in plant communities subject to recurrent disturbances. *Vegetatio* 43:5–21.

Novikoff, A. B. 1945. The concept of integrative levels and biology. *Science* 101:209–215.

Odum, E. P. 1953. *Fundamentals of Ecology.* W. B. Saunders, Philadelphia.

Odum, E. P. 1960. Organic production and turnover in old field succession. *Ecol.* 41:34–49.

Odum, E. P. 1969. The strategy of ecosystem development. *Science* 164:262–270.

Odum, E. P. 1971. *Fundamentals of Ecology* (3rd ed.). W. B. Saunders, Philadelphia.

Odum, E. P. 1977. The emergence of ecology as a new integrative discipline. *Science* 195:1289–1293.

Odum, H. T. 1957. Trophic structure and productivity of Silver Springs, Florida. *Ecol. Monogr.* 27:55–112.

O'Neill, R. V. 1971. Systems approaches to the study of forest floor arthropods. In *Systems Analysis and Simulation in Ecology*, ed. B. C. Patten, Volume I, pp. 441–477. Academic Press, New York.

O'Neill, R. V. 1976. Ecosystem persistence and heterotrophic regulation. *Ecol.* 57:1244–1253.

O'Neill, R. V. 1979a. Transmutations across hierarchical levels. In *Systems Analysis of Ecosystems*, ed. G. S. Innis and R. V. O'Neill, pp. 59–78. Int. Coop. Publ. House, Fairland, MD.

O'Neill, R. V. 1979b. A review of linear compartmental analysis in ecosystem science. In *Compartmental Analysis of Ecosystem Models*, ed. J. H. Matis, B. C. Patten, and G. C. White pp. 3–27. Int. Coop. Publ. House, Fairland.

O'Neill, R. V., and J. M. Giddings. 1979. Population interactions and ecosystem function: phytoplankton competition and community production. In *Systems Analysis of Ecosystems*, ed. G. S. Innis and R. V. O'Neill, pp. 103–123. Int. Coop. Publ. House, Fairland.

O'Neill, R. V., and D. E. Reichle. 1980. Dimensions of ecosystem theory. In *Forests: Fresh Perspectives from Ecosystem Analysis*, ed. R. H. Warinig, pp. 11–26. Oregon State Univ. Press, Corvallis.

O'Neill, R. V., and J. B. Waide. 1981. Ecosystem theory and

the unexpected: implications for environmental toxicology. *Management of Toxic Substances in Our Ecosystems*, ed. B. W. Cornaby, pp. 48–73. Ann Arbor Science, Ann Arbor.

O'Neill, R. V., R. H. Gardner, and D. E. Weller. 1982. Chaotic models as representations of ecological systems. *Amer. Nat.* 120:259–263.

Onsager, L. 1931. Reciprocal relations in irreversible processes, I. *Phys. Rev.* 37:405–426.

Orians, G. H., and O. T. Solbrig., eds. 1977. *Convergent Evolution in Warm Deserts*. Hutchinson Ross, Stroudsburg, Pa.

Oster, G. 1974. Stochastic behavior of deterministic models, pp. 24–37. In *Ecosystem Analysis and Prediction*, ed. S. A. Levin, pp. 24–37. Society for Industrial and Applied Mathematics, Philadelphia.

Oster, G., and Y. Takahashi. 1974. Models for age-specific interactions in a periodic environment. *Ecol. Monogr.* 44:483–501.

Overton, W. S. 1972. Toward a general model structure for forest ecosystems. In *Proc. Symp. on Research on Coniferous Forest Ecosystems*, ed. J. F. Franklin, Northwest Forest Range Station, Portland.

Overton, W. S. 1974. Decomposabilitty: a unifiying concept? *Ecosystem Analysis and Prediction*, ed. S. A. Levin, pp. 297–298. Society for Industrial and Applied Mathematics, Philadelphia.

Overton, W. S. 1975. The ecosystem modeling approach in the coniferous forest biome. In *Systems Analysis and Simulation in Ecology*, ed. B. C. Patten, Volume III, pp. 117–138. Academic Press, New York.

Overton, W. S. 1977. A strategy of model construction. In *Ecosystem Modelling in Theory and Practice: An Introduction with Case Histories*, ed. C.A.S. Hall and J. W. Day, pp. 50–73. John Wiley and Sons, New York.

Paine, R. T. 1966. Food web complexity and species diversity. *Amer. Nat.* 100:65–75.

Paine, R. T. 1969. The *Pisaster–Tegula* interaction: prey patches, predator food preference, and intertidal community structure. *Ecol.* 50:950–961.

Paine, R. T. 1974. Intertidal community structure: experimental studies on the relationship between a dominant competitor and its principal predator. *Oecologia* 15:93–120.

Paine, R. T. 1980. Food webs: linkage, interaction strength, and community infrastructure. *J. Anim. Ecol.* 49:667–685.

Paine, R. T., and S. A. Levin. 1981. Intertidal landscapes: disturbance and the dynamics of pattern. *Ecol. Monogr.* 51:145–178.

Parsons, D. J., and A. R. Moldenke. 1975. Convergence in vegetation structure along analogous climatic gradients in California and Chile. *Ecol.* 56:950–957.

Pattee, H. H. 1969. Physical conditions for primitive functional hierarchies. In *Hierarchical Structures*, ed. L. L. Whyte, A. G. Wilson, and D. Wilson, pp. 161–177. Elsevier, New York.

Pattee, H. H. 1972. The evolution of self-simplifying systems. In *The Relevance of General Systems Theory*, ed. E. Lazlo, pp. 31–42. Braziller, New York.

Pattee, H. H. 1978. The complementarity principle in biological and social structures. *J. Social Biol. Struct.* 1:191–200.

Pattee, H. H., ed. 1973. *Hierarchy Theory*. Braziller, New York.

Patten, B. C. 1959. An introduction to the cybernetics of the ecosystem: the trophic dynamic aspect. *Ecol.* 40:221–231.

Patten, B. C. 1975. Ecosystem linearization: an evolutionary design problem. In *Ecosystem Analysis and Prediction*, ed.

S. A. Levin, pp. 182–202. Society for Industrial and Applied Mathematics, Philadelphia.

Patten, B. C. 1978. Systems approach to the concept of the environment. *Ohio J. Sci.* 78:206–222.

Patten, B. C. 1982. Environs: relativisitic elementary particles for ecology. *Amer. Nat.* 119:179–219.

Patten, B. C., R. W. Bosserman, J. T. Finn, and W. G. Cale. 1976. Propagation of cause in ecosystems. In *Systems Analysis and Simulation in Ecology*, ed. B. C. Patten, Volume IV, pp. 457–579. Academic Press, New York.

Patten, B. C., and E. P. Odum. 1981. The cybernetic nature of ecosystems. *Amer. Nat.* 118:886–895.

Peters, R. H. 1976. Tautology in evolution and ecology. *Amer. Nat.* 110:1–12.

Peterson, R. O., R. E. Page, and K. M. Dodge. 1984. Wolves, moose, and the allometry of population cycles. *Science* 224:1350–1352.

Phillips, J. 1931. The biotic community. *J. Ecol.* 19:1–24.

Phillips, J. 1934. Succession, development, the climax, and the complex organism. *J. Ecol.* 22:554–571.

Phillips, J. 1935. Succession, development, the climax, and the complex organism, Part 2. *J. Ecol.* 23:210–246.

Pianka, E. R. 1974. *Evolutionary Ecology*. Harper and Row, New York.

Pielou, E. C. 1972. Niche width and niche overlap: a method for measuring them. *Ecol.* 53:687–692.

Pielou, E. C. 1977. *Mathematical Ecology*. John Wiley and Sons, New York.

Pimm, S. L. 1979. The structure of food webs. *Theor. Pop. Biol.* 16:144–158.

Pimm, S. L. 1980. Food web design and the effects of species deletion. *Oikos* 35:139–149.

Pimm, S. L. 1982. *Food Webs*. Chapman and Hall, London.

Pimm, S. L., and J. H. Lawton. 1977. Number of trophic levels in ecological communities. *Nature* 268:329–331.

Pimm, S. L., and J. H. Lawton. 1978. On feeding on more than one trophic level. *Nature* 275:542–544.

Pimm, S. L., and J. H. Lawton. 1980. Are food webs divided into compartments? *J. Anim. Ecol.* 49:879–898.

Plant, R. E., and M. Kim. 1975. On the mechanism underlying bursting in the Aplysia abdominal ganglion R12 cell. *Math. Biosci.* 26:357–375.

Plant, R. E., and M. Kim. 1976. Mathematical description of a bursting pacemaker neuron by a modification of the Hodgkin–Huxley equations. *Biophysical J.* 16:227–244.

Platt, J. R. 1969. Theories on boundaries in hierarchical systems. In *Hierarchical Structures*, ed. L. L. Whyte, A. G. Wilson, and D. Wilson, pp. 201–213. Elsevier, New York.

Platt, T., and K. L. Denman. 1975. Spectral analysis in ecology. *Ann. Rev. Ecol. Syst.* 6:189–210.

Pomeroy, L. R. 1970. The strategy of mineral cycling. *Ann. Rev. Ecol. Syst.* 1:171–190.

Post, W. M., and S. L. Pimm. 1983. Community assembly and food web stability. *Math. Biosci.* 64:169–192.

Poulson, T. L., and W. B. White. 1969. The cave environment. *Science* 165:971–980.

Prigogine, I. 1945. Modération et transformations irréversibles des systèmes ouverts. *Bull. Acad. Roy. Belg. Cl. Sci.* 31:600.

Prigogine, I. 1947. *Etude thermodynamique des processus irréversibles*. Desoer, Liege.

Prigogine, I., G. Nicholis, and A. Babloyantz. 1972a. Thermodynamics of evolution I. *Physics Today* 25(11):23–28.

Prigogine, I., G. Nicholis, and A. Babloyantz. 1972b. Thermodynamics of evolution II. *Physics Today* 25(12):38–44.

Quinlin, A. V. 1975. Design and analysis of mass conservative models of ecodynamic systems. Ph.D. diss., MIT, Cambridge.

Ramensky, L. G. 1924. Vestnik opytnozo dela Sredne-Chernoz Obl, Voronezh 37–73. Excerpt in *Readings in Ecology*, ed. E. J. Kormondy, pp. 152–162. Prentice-Hall, Englewood Cliffs, N.J.

Raunkiaer, C. 1934. *The Life Forms of Plants and Statistical Plant Geography: Being the Collected Papers of C. Raunkiaer*. Oxford Univ. Press, Oxford.

Regal, P. J. 1977. Ecology and evolution of flowering plant dominance. *Science* 196:622–629.

Regier, H. A., and D. J. Rapport. 1978. Ecological paradigms, once again. *Bull. Ecol. Soc. of Amer.* 59(1):2–6.

Reichle, D. E., R. V. O'Neill, and W. F. Harris. 1975. Principles of energy and material exchange in ecosystems. In *Unifying Concepts in Ecology*, ed. W. H. von Dobben and R. H. Lowe-McConnell, pp. 27–43. Dr. W. Junk, The Hague.

Rigler, F. H. 1975. The concept of energy flow and nutrient flow between trophic levels, pp. 15–26. In *Unifying Concepts in Ecology*, ed. W. H. von Dobben and R. H. Lowe-McConnell, pp. 15–26. Dr. W. Junk, The Hague.

Roberts, A., and K. Tregonning. 1980. The robustness of natural systems. *Nature* 288:265–266.

Roberts, G. F., and F. DiCesare. 1982. A systems engineering methodology for structuring and calibrating lake ecosystem models. *IEEE Trans. Systems, Man and Cybernetics* 12:3–14.

Romme, W. H. 1982. Fire and landscape diversity in subalpine forests of Yellowstone National Park. *Ecol. Monogr.* 52:199–221.

Root, R. B. 1967. The niche exploitation pattern of the blue-gray gnatcatcher. *Ecol. Monogr.* 37:317–350.

Root, R. B. 1975. Some consequences of ecosystem texture. In *Ecosystem Analysis and Prediction*, ed. S. A. Levin, pp. 83–92. Society for Industrial and Applied Mathematics, Philadelphia.

Rosen, R. 1972. Some systems theoretical problems in biology. In *The Relevance of General Systems Theory*, ed. E. Lazlo, pp. 43–66. Braziller, New York.

Rosen, R. 1977a. Complexity as a system property. *Gen. Sys.* 3:227–232.

Rosen, R. 1977b. Observation and biological systems. *Bull. Math. Biol.* 39:663–678.

Rowe, J. S. 1961. The level-of-integration concept and ecology. *Ecol.* 42:420–427.

Sage, R. D. 1973. Convergence of the lizard faunas of the chaparral habitats in central Chile and California. In *Mediterranean Type Ecosystems: Origin and Structure*, ed. F. Di Castri and H. A. Mooney, pp. 339–348. Springer-Verlag, Heidelberg.

Sakamoto, M. 1966. Primary production by phytoplankton community in some Japanese lakes and its dependence on lake depth. *Arch. Hydrobiol.* 62:1–28.

Saunders, P. T., and M. W. Ho. 1976. On the increase in complexity in evolution. *J. Theor. Biol.* 63:375–384.

Scavia, D. 1980. Conceptual model of phosphorus cycling. In *Nutrient Cycling in the Great Lakes*, ed. D. Scavia and R. Moll, pp. 119–140. Special Report 83, Great Lakes Research Division, University of Michigan, Ann Arbor.

Schindler, D. W. 1977. Evolution of phosphorus limitation in lakes. *Science* 195:260–262.

Schindler, D. W. 1978. Factors regulating phytoplankton production in the world's freshwaters. *Limnol. Oceanogr.* 23:478–486.

Schindler, J. E., J. B. Waide, M. C. Waldron, J. J. Hains, S. P. Schreiner, M. L. Freedman, S. L. Benz, D. R. Pettigrew, L. A. Schissel, and P. J. Clark. 1980. A microcosm approach to the study of biogeochemical systems, I: Theoretical rationale. In *Microcosms in Ecological Research*, ed. J. P. Giesy, pp. 192–203. Technical Information Center, U.S. Department of Energy, Symposium Series 52 (CONF-781101).

Schoffeniels, E. 1976. *Anti-Chance*. Pergamon Press, New York.

Schrodinger, E. 1945. *What Is Life?* Cambridge Univ. Press, Cambridge.

Schultz, A. M. 1967. The ecosystem as a conceptual tool in the management of natural resources. In *Natural Resources: Quality and Quantity*, ed. S. V. Cieriacy-Wantrup, pp. 139–161. Univ. of California Press, Berkeley.

Schultz, A. M. 1969. A study of an ecosystem: the arctic tundra. In *The Ecosystem Concept in Natural Resource Management*, ed. G. M. Van Dyne, pp. 73–93. Academic Press, New York.

Schumm, S. A., and R. W. Lichty. 1965. Time, space, and causality in geomorphology. *Amer. J. Sci.* 263:110–119.

Schuster, P., and K. Sigmund 1980. A mathematical model of the hypercycle. In *Dynamics of Synergeticis*, ed. H. Haken, pp. 170–179. Springer-Verlag, Heidelberg.

Sernander, R. 1908. On the evidences of postglacial changes of climate furnished by the peat mosses of northern Europe. *Geologiska Foreningens Forhandlingar* 30:465–473.

Shelford, V. E. 1913. *Animal Communities in Temperate America*. Univ. of Chicago Press, Chicago.

Shelford, V. E. 1931. Some concepts of Bioecology. *Ecol.* 12:455–467.

Shilov, I. A. 1981. The biosphere, levels of organization of life, and problems of ecology. *Soviet J. Ecol.* 12(1):1–6.

Shimwell, D. W. 1971. *The Description and Classification of Vegetation*. Sedgwick and Jackson, London.

Shugart, H. H., and D. C. West. 1981. Long-term dynamics of forest ecosystems. *Amer. Sci.* 69:647–652.

Siljak, D. D. 1975. When is a complex ecosystem stable? *Math. Biosci.* 25:25–50.

Simberloff, D. S. 1976. Trophic structure determination and equilibrium in an arthropod community. *Ecol.* 57:395–398.

Simberloff, D. S. 1981. Community effects of introduced species. In *Biotic Crisis in Ecological and Evolutionary Time*, ed. M. H. Nitecki, pp. 53–83. Academic Press. New York.

Simon, H. A. 1962. The architecture of complexiity. *Proc. Amer. Phil. Soc.* 106:467–482.

Simon, H. A. 1969. *The Sciences of the Artificial.* MIT Press, Cambridge.

Simon, H. A. 1973. The organization of complex systems. In *Hierarchy Theory*, ed. H. H. Pattee pp. 3–27. Braziller, New York.

Sjors, H. 1955. Remarks on ecosystems. *Svensk Botanisk Tidskrift* 49:155–169.

Slobodkin, L. B., F. E. Smith, and N. G. Hairston. 1967. Regulation in terrestrial ecosystems and the implied balance of nature. *Amer. Nat.* 101:109–124.

Smith, F. E. 1975. Ecosystems and evolution. *Bull. Ecol. Soc. of Amer.* 56(4):2–6.

Smith, R. L. 1966. *Ecology and Field Biology.* Harper and Row, New York.

Smith, V. H. 1982. The nitrogen and phosphorus dependence of algal biomass in lakes: an empirical and theoretical analysis. *Limnol. Oceanogr.* 27:1101–1112.

Smuts, J. C. 1926. *Holism and Evolution.* Macmillan, New York.

Sollins, P., G. Spycher, and C. Topik. 1983. Processes of soil organic-matter accretion at a mudflow chronosequence, Mt. Shasta, California. *Ecol.* 64:1273–1282.

Sousa, W. P. 1979. Experimental investigation of disturbance and ecological succession in a rocky intertidal algal community. *Ecol. Monogr.* 49:227–254.

Southwood, T.R.E., R. M. May, M. P. Hassell, and G. R.

Conway. 1974. Ecological strategies and population parameters. *Amer. Nat.* 108:791–804.

Sprugel, D. G. 1976. Dynamic structure of wave-regenerated *Abies balsamea* forests in the north-eastern United States. *J. Ecol.* 64:889–911.

Sprugel, D. G., and F. H. Bormann. 1981. Natural disturbance and the steady state in high-altitude balsam fir forests. *Science* 211:390–393.

Stanley, S. M. 1981. *The New Evolutionary Timetable.* Basic Books, New York.

Stent, G. S. 1978. *Paradoxes of Progress.* W. H. Freeman, San Francisco.

Stommel, H. 1963. Varieties of oceanographic experience. *Science* 139:572–576.

Strong, D. R. 1982. Null hypotheses in ecology. In *Conceptual Issues in Ecology*, ed. E. Saarinen, pp. 245–259. Reidel, Boston.

Summerhagen, V. S., and C. S. Elton. 1923. Contribution to the ecology of Spitsbergen and Bear Island. *J. Ecol.* 11:214–286.

Tansky, M. 1978. Stability of multispecies predator–prey systems. *Memoirs Coll. Sci., Univ. Kyoto, Ser. B.* 7(2):87–94.

Tansley, A. G. 1935. The use and abuse of vegetational concepts and terms. *Ecol.* 16:284–307.

Taylor, W. P. 1935a. Some animal relations to soil. *Ecol.* 16:127–136.

Taylor, W. P. 1935b. Significance of the biotic community in ecological studies. *Quart. Rev. Biol.* 10:291–307.

Taylor, W. P. 1936. What is ecology and what good is it? *Ecol.* 17:333–346.

Tregonning, K., and A. Roberts. 1979. Complex systems which evolve toward homeostasis. *Nature* 281:563–564.

Ulanowicz, R. E. 1979. Prediction, chaos, and ecological

perspective. In *Theoretical Systems Ecology*, ed. E. A. Halfon, pp. 107–117. Academic Press, New York.

Ulanowicz, R. E. 1983. Identifying thte structure of cycling in ecosystems. *Math. Biosci.* 65:219–237.

Van Voris, P., R. V. O'Neill, W. R. Emanuel, and H. H. Shugart. 1980. Functional complexity and ecosystem stability. *Ecol.* 61:1352–1360.

Vermeij, G. J. 1978. *Biogeography and Adaptation: Patterns in Marine Life*. Harvard Univ. Press, Cambridge.

Vollenweider, R. A. 1975. Input–output models with special reference to the phosphorus loading concept in limnology. *Schweiz. Z. Hydrol.* 37:53–84.

Vollenweider, R. A. 1976. Advances in defining critical loading levels for phosphorus in lake eutrophication. *Mem. Ist. Ital. Idrobiol.* 33:53–83.

Volterra, V. 1926. Variazioni e fluttuazioni del numero d'individui in specie animali conviventi. *Mem. Acad. Lincei.* 2:31–113. Translated in Chapman, R. N. 1931. *Animal Ecology*. McGraw-Hill, New York.

Waide, J. B., J. E. Krebs, S. P. Clarkson, and E. M. Setzler. 1974. A linear systems analysis of the calcium cycle in a forested watershed ecosystem. *Prog. in Theor. Biol.* 3:261–345.

Waide, J. B., J. E. Schindler, M. C. Waldron, J. J. Hains, S. P. Schreiner, M. L. Freedman, S. L. Benz. D. R. Pettigrew, L. A. Schissel, and P. J. Clark. 1980. A microcosm approach to the study of biogeochemical systems, 2: Responses of aquatic microcosms to physical, chemical, and biological perturbations. In *Microcosms in Ecological Research*, ed. J. P. Giesy, pp. 204–223. Technical Information Center, U.S. Department of Energy, Symposium Series 52 (CONF-781101).

Waide, J. B., and J. R. Webster. 1976. Engineering systems analysis: applicability to ecosystems. In *Systems Analysis and Simulation in Ecology*, ed. B. C. Patten, Volume IV, pp. 329–371. Academic Press, New York.

Walter, E. 1982. *Identifiability of State Space Models.* Lecture Notes in Biomathematics 46. Springer-Verlag, New York.

Waring, R. H., and J. F. Franklin. 1979. Evergreen coniferous forests of the Pacific Northwest. *Science* 204:1380–1386.

Warming, E. 1895. *Plantesamfund Grundtrak of den okologiska Plantegeographi.* Copenhagen.

Warming, E. 1909. *Oecology of Plants: An Introduction to the Study of Plant Communities.* Translated by P. Groom and I. B. Balfour. Oxford Univ. Press, Oxford.

Watt, A. S. 1925. On the ecology of British beechwoods with special reference to their regeneration. *J. Ecol.* 13:27–73.

Watt, A. S. 1947. Pattern and process in the plant community. *J. Ecol.* 35:1–22.

Watt, K.E.F. 1969. A comparative study on the meaning of stability in five biological systems: insect and furbearer populations, influenza, Thai hemorrhagic fever, and plague. In *Diversity and Stability in Ecological Systems*, ed. G. M. Woodwell and H. H. Smith, pp. 142–150. Brookhaven Symposium in Biology 22. U.S. Department of Commerce, Springfield, MD.

Weaver, W. 1948. Science and complexity. *Amer. Sci.* 36:537–544.

Webb, L. J. 1968. Environmental relationships of the structural types of Australian rain forest vegetation. *Ecol.* 49:296–311.

Webb, W. L., W. K. Lauenroth, S. R. Szarek, and R. S. Kinerson, 1983. Primary production and abiotic control in forests, grasslands, and desert ecosystems in the United States. *Ecol.* 64:134–151.

Webster, J. R. 1979. Hierarchical organization of ecosystems. In *Theoretical Systems Ecology*, ed. E. Halfon, pp. 119–131. Academic Press, New York.

Weinberg, G. M. 1975. *An Introduction to General Systems Thinking*. John Wiley and Sons New York.

Weinberg, G. M., and D. Weinberg. 1979. *On the Design of Stable Systems*. John Wiley and Sons, New York.

Weiss, P. A., ed. 1971. *Hierarchically Organized Systems in Theory and Practice*. Hafner, New York.

White, C., and W. S. Overton. 1974. *Users Manual for the FLEX2 and FLEX3 Model Processors*. Bulletin 15, Forest Research Laboratory, Oregon State Univ., Corvallis.

White, P. S. 1979. Pattern, process, and natural disturbance in vegetation. *Bot. Rev.* 45:229–299.

Whitehead, D. R. 1964. Fossil pine pollen and full glacial vegetation in southeastern North Carolina. *Ecol.* 44:403–406.

Whitehead, D. R. 1965. Palynology and pleistocene phytogeography of unglaciated eastern North America. In *Quaternary of the United States*, ed. H. E. Wright and D. G. Frey, pp. 417–451. Princeton Univ. Press, Princeton.

Whittaker, R. H. 1953. A consideration of climax theory: the climax as a population pattern. *Ecol. Monogr.* 23:41–78.

Whittaker, R. H. 1956. Vegetation of the Great Smoky Mountains. *Ecol. Monogr.* 26:1–80.

Whittaker, R. H. 1962. Classification of natural communities. *Bot. Rev.* 28:1–239.

Whittaker, R. H. 1967. Gradient analysis of vegetation. *Biol. Rev.* 42:207–264.

Whittaker, R. H. 1975. *Communities and Ecosystems*. Macmillan, New York.

Whittaker, R. H., and G. M. Woodwell. 1972. Evolution of natural communities. In *Ecosystem Structure and Func-*

tion, ed. J. A. Wiens, pp. 137–159. Oregon State Univ. Press, Corvallis.

Whyte, L. L. 1969. Structural hierarchies: a challenging class of physical and biological problems. In *Hierarchical Structures*, ed. L. L. Whyte, A. G. Wilson, and D. Wilson, pp. 3–16. Elsevier, New York.

Whyte, L. L., A. G. Wilson, and D. Wilson, eds. 1969. *Hierarchical Structures*. Elsevier, New York.

Wicken, J. S. 1980. A thermodynamic theory of evolution. *J. Theor. Biol.* 87:9–23.

Wicken, J. S. 1985. Holism and hierarchy: A thermodynamic theory of evolution. Manuscript.

Wiegert, R. G. 1974. A general ecological model and its use in simulating algal–fly energetics in a thermal spring community. In *Insects: Studies in Population Management*, ed. P. W. Geier, L. R. Clark, D. J. Anderson, and H. A. Nix, pp. 85–102. Ecol. Soc. of Australia, Canberra.

Wiegert, R. G. 1975. Simulation modeling of the algae–fly components of a thermal ecosystem. In *Systems Analysis and Simulation in Ecology*, Volume III, ed. B. C. Patten, pp. 157–181. Academic Press, New York.

Wiegert, R. G., and D. F. Owen. 1972. Trophic structure, available resources, and population density in terrestrial vs. aquatic ecosystems. *J. Theor. Biol.* 30:69–81.

Wilde, S. A. 1968. Mycorrhizae and tree nutrition. *Bioscience* 18:482–484.

Wilson, E. O. 1969. The species equilibrium. In *Diversity and Stability in Ecological Systems*, ed. G. M. Woodwell and H. H. Smith, pp. 38–47. U.S. Department of Commerce, Springfield.

Woodwell, G. M. and H. H. Smith eds., 1969. *Diversity and Stability in Ecological Systems*. Brookhaven Symposium in Biology 22. U.S. Department of Commerce, Springfield.

Wright, R. F. 1974. Forest fire: impact on the hydrology,

chemistry, and sediments of small lakes in northeastern Minnesota. Interim Report No. 10, Limnology Research Center, Univ. of Minnesota, Minneapolis.

Wright, S. 1959. Genetics and the hierarchy of biological sciences. *Science* 130:959–965.

Zackrisson, O. 1977. Influence of forest fires on the North Swedish boreal forest. *Oikos* 29:22–32.

Zaret, T. M. 1982. The stability/diversity controversy: a test of hypotheses. *Ecol.* 63:721–731.

Zedler, P. H., C. R. Gautier, and G. S. McMaster. 1983. Vegetation change in response to extreme events: the effect of a short interval between fires in California chaparral and coastal shrub. *Ecol.* 64:809–818.

Author Index

247

248

Subject Index

251

Library of Congress Cataloging-in-Publication Data

A hierarchical concept of ecosystems.

(Monographs in population biology ; 23)
Bibliography: p.
Includes indexes.
1. Ecology.   2. Biotic communities.   I. O'Neill, R. V.
(Robert V.), 1940- .   II. Series.
QH541.H525   1986   574.5   86-9423
ISBN 0-691-08436-X (alk. paper)
ISBN 0-691-08437-8 (pbk.)

R. V. O'Neill and D. L. DeAngelis are Senior Ecologists at Oak
Ridge National Laboratory in Oak Ridge, Tennessee. J. B. Waide
is Research Ecologist at Coweeta Hydrologic Laboratory in Otto,
North Carolina. T.F.H. Allen is Professor of Botany at the University of Wisconsin, Madison.